Physics at Surfaces and Interfaces

Physics at Surfaces and Interfaces

Proceedings of the International Conference

Puri, India 4 - 8 March 2002

Editor

B. N. Dev

Institute of Physics, India

World Scientific

New Jersey • London • Singapore • Hong Kong

Published by

World Scientific Publishing Co. Pte. Ltd.
5 Toh Tuck Link, Singapore 596224
USA office: Suite 202, 1060 Main Street, River Edge, NJ 07661
UK office: 57 Shelton Street, Covent Garden, London WC2H 9HE

British Library Cataloguing-in-Publication Data
A catalogue record for this book is available from the British Library.

Proceedings of the International Conference on
PHYSICS AT SURFACES AND INTERFACES

ISBN 981-238-575-4

Printed in Singapore by World Scientific Printers (S) Pte Ltd

International Avisory Committee

G. S. Agarwal	India
D. D. Bhawalkar	India
D. Chakravorty	India
R. Chidambaram	India
R. K. Choudhury	India
M. P. Das	Australia
B. A. Dasannacharya	India
H. Dosch	Germany
S. Duttagupta	India
S. S. Jha	India
S. K. Joshi	India
P. K. Kaw	India
N. Kumar	India
Vikram Kumar	India
K. Lal	India
R. A. Mashelkar	India
G. K. Mehta	India
A. Nigavekar	India
B. Raj	India
C. N. R. Rao	India
T. V. Ramakrishnan	India
V. S. Ramamurthy	India
B. B. Rath	USA
J. R. Schneider	Germany
Bikash Sinha	India
S. K. Sinha	USA
K. Yamashita	Japan

Local Organizing Committee

R. K. Choudhury (Chairman)
B. N. Dev (Convener)
S. M. Bhattacharjee
A. M. Jaynnavar
D. P. Mahapatra
S. N. Sahu
P. V. Satyam
B. R. Sekhar
T. Som
S. Varma

Preface

This book is a collection of selected papers, mostly invited, presented at the International Conference on Physics at Surfaces and Interfaces (PSI2002). The conference was held at Toshali Sands, Puri, India from March 4 to March 8, 2002 and organized by Institute of Physics, Bhubaneswar, India.

The conference had presentations on experimental and theoretical aspects of (a) growth on surfaces, (b) multilayers, (c) structural, electronic and magnetic properties, (d) semiconductor heterostructures, (e) quantum devices, (f) catalysis, (g) surface and interface modifications by ion-solid interactions and (h) new techniques.

Some invited talks presented in the conference are not included in this volume due to the publication of the material elsewhere or other commitments of the speakers. However, they had an important impact on discussions. Interested readers are encouraged to consult appropriate references. For this purpose these talks are listed here. Contributed oral and poster presentations are also listed.

I wish to acknowledge the organizations and individuals whose contributions made the successful organization of the conference and these Proceedings possible. I express my gratitude to the host institution and other sponsors: Board of Research in Nuclear Sciences (BRNS), Mumbai and Department of Science and Technology (DST), New Delhi. The support from Mack International, Mumbai, is acknowledged.

Particular thanks go to the director of Institute of Physics, Bhubaneswar, R. K. Choudhury, for providing suggestions and help in organizing the conference. My thanks are due to other members of the local organizing committee. My special thanks go to those doctoral students who provided their help, especially D. K. Goswami, B. Satpati and K. Bhattacharjee, throughout including the preparation of this volume. D. Pradhan has provided secretarial assistance in organizing the conference. Suggestions from the members of the International Advisory Committee have been an important contribution.

B. N. Dev

Unpublished Papers

Invited Papers

1. Universality in the Orientational Phase Diagram of Vicinal Surfaces
   S. M. Bhattacharjee

2. Strained Quantum Structures: Wells and Dots, Synthesis and Spectroscopy
   B. M. Arora

3. X-Ray Investigation of the Nanoscale Structure of Liquid Surfaces and Interfaces
   J. Daillant

4. Novel Superconducting Phenomena at Surfaces and Interfaces
   M. P. Das

5. Important New Analytical Opportunities Based on Compact X-Ray Source-Optics Combinations
   W. M. Gibson

6. Electron Microscopy of Semiconductor and Oxide Surfaces
   L. D. Marks

7. From X-Ray Standing Waves to Real Space Imaging of Atoms
   G. Materlik

8. Surface and Interface Studies Using Indus-1
   R. V. Nandedkar

9. Physics on a Quantum Cascade Laser
   T. Chakraborty

10. Collaborative Research: Need of the Times
    B. A. Dasannacharya

Oral Contributory Papers

Poster Papers

Introduction

Since early sixties there has been a tremendous growth in the study of solid surfaces and solid–solid interfaces. The interest has also been extended to solid–electrolyte interfaces and liquid surfaces and interfaces. Understanding various phenomena at surfaces and interfaces are not only of academic interest, surface and interface properties also influence many areas of applications, for example, lower dimensional structures in semiconductor devices. Quantum *well*, *wire* and *dot* structures and the related device physics are widely studied. These Proceedings dealing with the physics at surfaces and interfaces have been organized in four sections in the following scheme.

Section 1 deals with theoretical and experimental aspects of the structure and morphology involving clean surfaces and adsorbed layers on surfaces. The articles also deal with vacuum deposition, electrodeposition, catalytic behaviour and electronic structure.

Section 2 deals with the structure and transport in quantum structures. Semiconductor devices with quantum *well* structures, spin transport in 2DEG, metallic quantum *wire* on semiconductors and semiconductor quantum *dot* structures are the subject matter of this section.

Articles dealing with basic aspects and interdisciplinary applications of layered synthetic microstructures (LSMs) are presented in section 3. Analysis of interface roughness and layer composition of LSMs, fabrication of hard X-ray telescopes with LSMs for X-ray astronomy and the effect of diffusion across interfaces of such multilayer systems are presented here.

Interaction of energetic ion beams with solid surfaces lead to various kinds of surface modification. An understanding of such interactions has provided new surface analytical tools and has been useful in semiconductor device fabrication. Section 4 deals with semiconductor surfaces exposed to ion beams and ion irradiated semiconductor multilayers.

Contents

CLEAN SURFACES AND ADSORBED LAYERS: STRUCTURE AND MORPHOLOGY

HONEYCOMBS, TRIANGLES AND BRIGHT STARS: THE ADATOM-INDUCED RECONSTRUCTION OF PT(111)

SHOBHANA NARASIMHAN AND RAGHANI PUSHPA
Jawaharlal Nehru Centre for Advanced Scientific Research,
Jakkur PO, Bangalore 560064 India
E-mail: shobhana@jncasr.ac.in

We have studied the reconstruction of Pt(111) theoretically, using a two-dimensional Frenkel-Kontorova model whose parameters have been determined from *ab initio* calculations. In agreement with experiment, we find that the surface does not reconstruct under normal conditions. However, if the chemical potential is lowered, e.g., by the presence of adatoms, the surface reconstructs into a pattern of FCC and HCP domains, whose size and geometry is very sensitive to the parameters. As the surface layer densifies, the pattern of domains evolves smoothly from honeycombs to triangles to bright stars to a Moiré pattern.

1 Introduction

1.1 Surface Reconstructions

When a bulk crystal is cleaved, atoms near the surface lose neighbours, and as a result are no longer under equilibrium at their bulk-truncated positions. The resulting rearrangement of atoms near the surface may be small or drastic, depending on the nature of the material and the direction of the cut. When the periodicity and/or symmetry of the surface structure differs from that obtained by a simple bulk truncation, the surface is said to have reconstructed. If the surface reconstructs, the altered structure can have large implications for growth and chemical reactions taking place on the surface.

The (111) surfaces of face-centered-cubic (FCC) metals were thought to be immune to reconstruction, as their bulk-truncated structure consists of a close-packed atomic arrangement which was presumed to be stable. However, experiments have shown that Au(111) reconstructs into a complex herring-bone structure, and Pt(111) can be induced to reconstruct.

1.2 Experimental data on Pt(111)

Pt(111) is one of the most important systems in surface science, due to its importance as a catalyst in many commercially important syntheses. Therefore, any information about its structure and properties is of great interest. Experiments have shown that while the surface is unreconstructed under normal conditions of temperature and pressure, it can be induced to reconstruct either by heating it to 1330 K,[1] or by depositing a large number of adatoms on the surface, by placing the surface in a supersaturated Pt vapour.[2] In both cases, the reconstructed structure (observed by X-ray diffraction[1] and scanning tunneling microscopy (STM)[2]) consists of domains of FCC stacking alternating with faulted areas of hexagonal close packed (HCP) stacking. The FCC and HCP domains are separated by partial misfit dislocation domain walls where atoms sit at bridge sites; these domain walls form either a honeycomb network[1,2] or, on certain dense terraces, a pattern of interlocking wavy triangles.[3] The periodicity of these patterns varies from about 100 to 300 Å.

In this paper, we would like to confirm and understand these experimental results. In particular, we would like to understand the factors controlling the presence/absence of the reconstruction, and its geometry and spacing. Such questions assume greater importance in the light of work which has shown that such long-period reconstructions formed by networks of misfit dislocations can be used as templates for synthesizing nanostructures comprised of regularly spaced arrays of islands.[4,5]

2 Calculations

2.1 Frenkel-Kontorova Model

Reconstructions like that seen on Pt(111) result from a competition between three effects: (i) surface atoms would like to be closer together than the bulk spacing and thus increase the density in the surface layer (ii) it costs energy when surface atoms lose registry with the substrate (iii) in general, it costs energy to supply the extra atoms needed to increase the density in the top layer. These three effects are included in the Frenkel-Kontorova model,[6] where the Hamiltonian of the system is given by

$$H = \sum_i V_{ss}(l_i) + \sum_j V_{sb}(\mathbf{r}_j) + \Gamma N. \tag{1}$$

The three terms in the Hamiltonian correspond to the three factors listed above. V_{ss} is the interaction between nearest neighbour surface atoms which are connected by bonds of length l_i; it has a minimum at a value $l_i = b < a$, where a is the bulk nearest neighbour spacing. The effects of the layers below are contained in the substrate potential V_{sb}; this is an external potential in which the surface atoms j sit at positions $\mathbf{r}_j$. Finally, the last term accounts for the fact that the unreconstructed and reconstructed surfaces have different numbers of atoms per unit cell. N is the total number of atoms, and Γ is a chemical potential whose value changes when the additional atoms come from the bulk, step edges, adatoms, etc.

We obtain V_{ss}, V_{sb} and Γ from *ab initio* density functional theory calculations, as described below.

2.2 Details of Ab Initio Calculations

Ab Initio calculations were performed to obtain the parameters for the Frenkel-Kontorova model. These were carried out using the PWSCF package,[7] with ultrasoft pseudopotentials and a plane wave basis with a cut-off of 20 Ry. The cut-off was increased to 160 Ry for the augmentation charges associated with the use of the ultrasoft pseudopotential. Exchange-correlation interactions were treated using the Ceperley-Alder form of the local density approximation.[8] For surface calculations, a supercell consisting of 9 layers of atoms and 6 layers of vacuum was used. Bulk calculations were carried out using 110 k-points in the irreducible Brillouin zone (IBZ) of the FCC structure, and surface calculations were carried out using 27 k-points in the IBZ of the (1×1) surface cell. To determine the adatom adsorption energy, a larger (3×3) surface cell with a single adatom adsorbed at an FCC site was used; the number of k-points was reduced commensurately. We obtain the bulk lattice constant a_0 as 3.92 Å, which is in excellent agreement with experiment.

2.3 Determination of Surface-Surface Interactions

Fig. 1 shows our *ab initio* results for the variation in the surface stress σ of a slab of Pt atoms with (111) faces when the in-plane nearest-neighbor distance

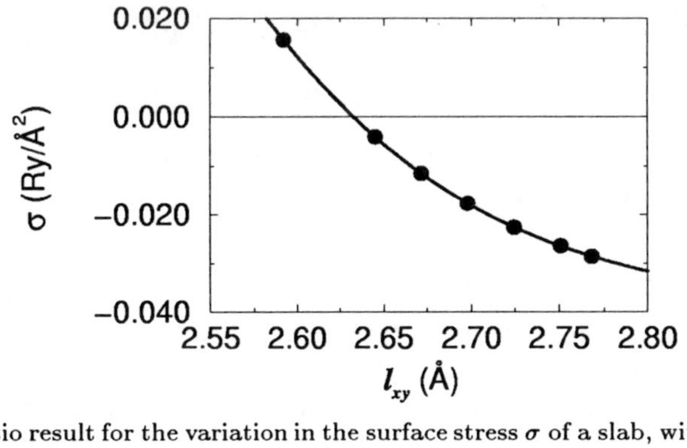

Figure 1. Ab initio result for the variation in the surface stress σ of a slab, with the in-plane nearest neighbor distance l_{xy}.

l_{xy} is varied. (For the bulk-truncated surface, $l_{xy} = a = 2.77\text{Å}$.) The convention used is that negative/positive σ corresponds to tensile/compressive stress. The first interlayer spacing d_{12} is relaxed for each l_{xy}, and it is therefore a good approximation to assume that the surface stress arises entirely from the bonds between surface atoms. One can then use the formula $\sigma = -(\sqrt{3}/l_{xy})(\partial V_{ss}/\partial l_{xy})$ to obtain the interaction V_{ss} between nearest-neighbour surface atoms. Upon approximating by a Morse potential $V_{ss} = A_0\{1 - \exp[-A_1(a-b)]\}^2$, we obtain $A_0 = 60.2$ mRy, $A_1 = 2.062$ Å^{-1}, and $b = 2.638$ Å; note in particular that $b < a$.

2.4 Determination of Surface-Bulk Interactions

In the Frenkel-Kontorova model, the effect of the bulk layers on the surface layer is described by an external potential V_{sb}, which contains information about the energy penalties when the surface layer loses registry with the substrate. We obtain these from *ab initio* computations of the change in the total energy of the slab when the surface layer occupies sites other than the most favoured FCC site. We find that the energy cost of these surface stacking faults for HCP, bridge and top sites is, respectively, $V_H = 5.05$ mRy, $V_B = 5.89$ mRy, and $V_T = 11.85$ mRy. The value of V_{sb} for an arbitrary position (x, y) of a surface atom is then given by expanding in a two-dimensional Fourier series about these values:

$$\begin{aligned} V_{sb}(x,y) &= (3V_B + V_T)/4 \\ &\quad - \left(\tfrac{V_H - 2V_T}{9}\right)\left\{\cos\left[\tfrac{4\pi}{\sqrt{3}a}y\right] + \cos\left[\tfrac{2\pi}{a}\left(x + \tfrac{y}{\sqrt{3}}\right)\right] + \cos\left[\tfrac{2\pi}{a}\left(x - \tfrac{y}{\sqrt{3}}\right)\right]\right\} \\ &\quad + \left(\tfrac{V_H + V_T}{9} - \tfrac{V_B}{4}\right)\left\{\cos\left[\tfrac{4\pi}{a}x\right] + \cos\left[\tfrac{2\pi}{a}(x + \sqrt{3}y)\right] + \cos\left[\tfrac{2\pi}{a}(x - \sqrt{3}y)\right]\right\} \\ &\quad + \left(\tfrac{V_H}{3\sqrt{3}}\right)\left\{\sin\left[\tfrac{2\pi}{a}\left(x + \tfrac{y}{\sqrt{3}}\right)\right] - \sin\left[\tfrac{4\pi}{\sqrt{3}a}y\right] - \sin\left[\tfrac{2\pi}{a}\left(x - \tfrac{y}{\sqrt{3}}\right)\right]\right\}, \end{aligned} \quad (2)$$

This expression has global minima (=0) at FCC sites, and local minima (= $V_H = 5.05$ mRy) at HCP sites.

2.5 Determination of the Chemical Potential

The last term in Eq. 1 involves the chemical potential Γ. This specifies the difference in energy between an atom sitting in the unreconstructed surface and an atom sitting at the site it occupies (bulk, step edge, adatom, etc.) before it is brought into the surface layer to increase the density of the surface. This quantity can be calculated from *ab initio* calculations of the surface energy, bulk cohesive energy, adatom adsorption energy and kink site adsorption energy. Accordingly, we obtain Γ_b = 50.56 mRy for bulk atoms, Γ_a = -67.12 mRy for adatoms, and Γ_k = 50.56 mRy for kink atoms.

2.6 Energy Minimization

Now that we have all the parameters for the Frenkel-Kontorova model, we find the optimal surface structure, as a function of the chemical potential Γ, by minimizing the surface energy $\gamma = E/A$, where E is the total energy given by Eq. 1, and A is the area of the surface. We consider two cases, that of uniaxial compression (along the $[1\bar{1}0]$ direction, $n+1$ surface atoms sit in n substrate spacings, so that the density is increased by $(n+1)/n$), and that of isotropic compression in the surface plane (the density is increased by a factor of $[(n+1)/n]^2$). For the latter case, we have an additional parameter θ, which specifies the angle between the unit vectors of the surface layer and the substrate. Fig. 2 shows our results for $\Delta\gamma$, the difference in the surface energies of the reconstructed and unreconstructed surfaces, as a function of the unit cell size n. We have shown results for the two extreme values of chemical potential, Γ_b (all extra atoms come from bulk sites) and Γ_a (all extra atoms come from adatoms). If the surface reconstructs, the curve of $\Delta\gamma$ vs. n will have a minimum at a finite value of n.

From Fig. 2, it is clear that the surface will not reconstruct if all extra atoms come from the bulk (or from step edges, since $\Gamma_k \approx \Gamma_b$.) However, if all extra atoms are from adatoms, the surface reconstructs, with the isotropic (threefold) reconstruction favoured over the uniaxial one. As Γ is increased

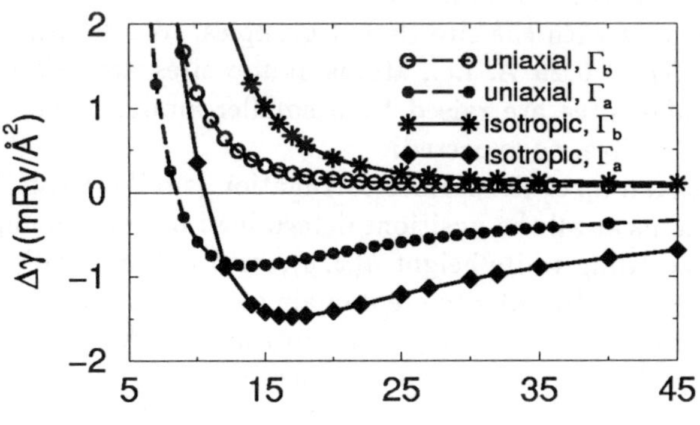

Figure 2. Ab initio result for the variation in the surface stress σ of a slab, with the in-plane nearest neighbor distance l_{xy}.

from Γ_a towards Γ_b (more of the additional density come from atoms initially in the bulk or at step edges, less from adatoms), n is increased. However once Γ crosses a threshold value of 36 mRy, it is no longer favourable for the surface to reconstruct.

These results agree with the experimental finding that Pt(111) does not reconstruct under normal conditions, but reconstructs into a pattern with threefold symmetry in the presence of a large number of adatoms (due to the presence of a supersaturated Pt vapour).

It should be noted that our "isotropic" structures consist of sixfold atomic coordination throughout; i.e., there are no point dislocations. Certain surfaces, such as Au(111), which reconstructs into a "herringbone" structure,[9,10] do display a regular array of point dislocations. However, for Pt(111) we find that structures containing point dislocations are not energetically favourable.

2.7 Domain Patterns

In order to compare with experimental STM images, and to get a clear picture of the periodicity and geometry of the arrangement of FCC and HCP domains, we use our theoretical results for the favoured atomic arrangement to simulate STM images. In order to do this, we obtain the atomic corrugation using a formula similar to Eq. 2, in which we replace V_H, V_B and V_T (the values of the substrate potential at HCP, bridge and top sites) by the corresponding heights h_H, h_B and h_T respectively, to obtain the height $h(x, y)$ of an atom at a general position (x, y) on the surface. The values of the heights at HCP, bridge and top sites, relative to the FCC sites, are obtained from our *ab initio* calculations on surface stacking faults, where we find that the height of the top layer changes with the site that it occupies. We get $h_H = 0.03$ Å, $h_B = 0.07$ Å, and $h_T = 0.29$ Å, i.e., atoms at top sites are raised the most, and atoms at bridge sites are raised by a smaller amount, relative to atoms at FCC and HCP sites, as expected.

To obtain a map of the surface corrugation (similar to an STM image), we then draw atoms at all the positions determined by minimizing Eq. 1, shading each atom according to its height $h(x, y)$, with lighter shading corresponding to a greater height. In order to see domain patterns clearly, we use a nonlinear grey scale, with white, light grey, dark grey and black corresponding to atoms at top, bridge, HCP and FCC sites respectively.

Fig. 3 shows how the domain patterns evolve as the density in the top layer is increased (i.e., n is decreased) for the isotropic case. Note that progressively smaller values of n are favoured by lower values of the chemical potential Γ. Also, as the density in the top layer is increased, we find that the angle of

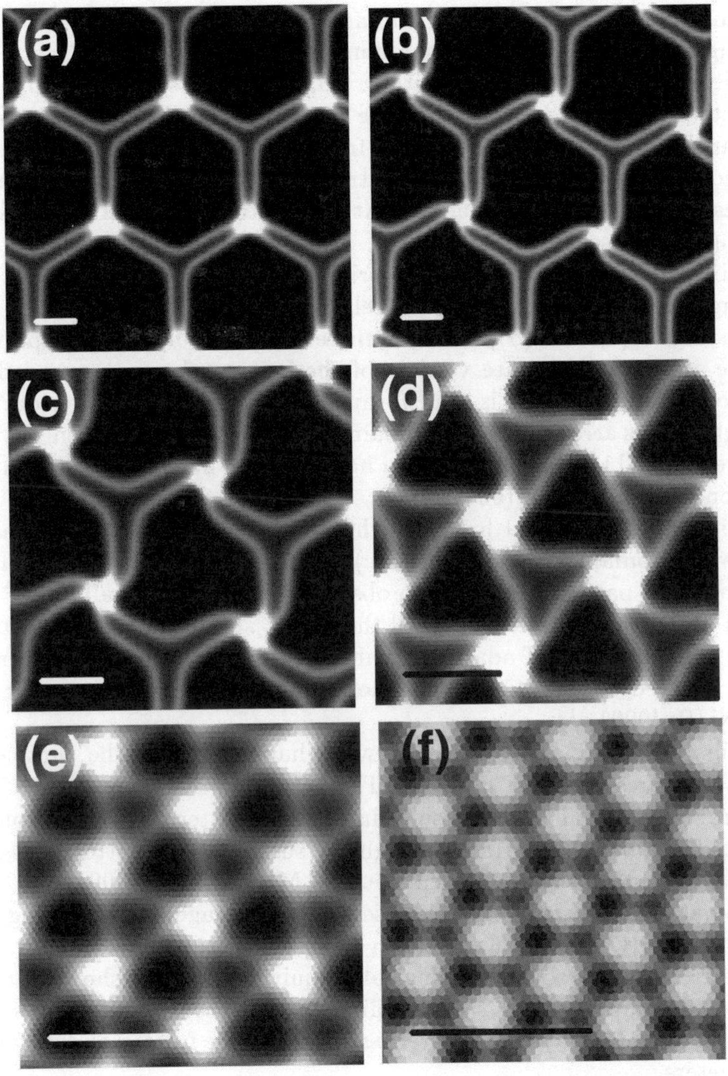

Figure 3. Simulated STM images of the Pt(111) surface, showing the evolution of the domain patterns as the density is increased (and n decreased) isotropically. The value of n is (a) 70 (b) 70 (c) 50 (d) 30 (e) 20 (f) 10. (a) and (b) differ in the value of θ, in (a) $\theta = 0$, whereas in (b) θ is optimized to 0.14°. The black/white line marked in each image has a length of 50 Å.

rotation θ increases, in an approximately linear manner.

Figs. 3(a) and 3(b) show the domain pattern for $n = 70$, corresponding to a density increase of 2.9%, with $\theta = 0$, and θ at its optimal value of 0.14°, respectively. It can be seen that the domain walls (bridge sites) separating the large hexagonal FCC domains from the narrow HCP domains form a honeycomb network that bears a striking similarity to experimental STM images of the Pt(111) surface.[2,3] At alternate vertices of the hexagons in Fig. 3(b), one sees "bright rotors". Atoms here sit at top sites, and the domain walls that stem from these bright areas have a twist that results from the slight rotation of the top layer relative to the substrate.

In Figs. 3(c) and (d), n has been reduced to 50 and 30 respectively. It can be seen that alternate vertices of the hexagon become rounded, so that the honeycomb network is transformed into a pattern of interlocking wavy triangles. This arrangement of domains has been seen experimentally on dense terraces of Pt(111).[3] Such a pattern has also been observed for Na/Au(111),[11] and for 3 ML of Cu/Ru(0001).[12]

Fig. 3(e) shows what happens when n is further reduced to a value of 20. The wavy domain walls of Figs. 3(c) and (d) straighten out, resulting in a "bright star" pattern. In order to observe this experimentally for Pt(111), it would be necessary that nearly all the excess surface density be provided by an initial deposition of adatoms, which is unlikely, and this pattern has not been observed for Pt(111). It has, however, been observed on other surfaces, e.g., Ni/Ru(0001).[13]

Finally, in Fig. 3(f), we show what happens when the surface layer is highly compressed, with $n = 10$. This situation is highly energetically unfavourable for Pt(111), and results in a Moiré pattern, with the compression distributed almost uniformly through the surface. This pattern has also been observed on other surface systems, e.g., 4 ML of Cu/Ru(0001).[12]

What is noteworthy about the sequence of images shown in Fig. 3 is that there is clearly a smooth transition between the different domain patterns, which are all seen to be topologically equivalent. This shows that the variety of domain patterns seen on different surface systems arises not from any fundamental physical differences, but merely from a slight variation in parameters.

3 Summary

Using a Frenkel-Kontorova model parametrized entirely by *ab initio* calculations, we have studied the reconstruction of Pt(111). In agreement with experiment, we find that the surface does not reconstruct under normal conditions. However, if the chemical potential is lowered slightly (as happens in the presence of a large number of adatoms, when the surface is placed in a supersaturated Pt vapour), the surface reconstructs, with a periodicity and geometry determined by the chemical potential.

In fact, one can show by a stability analysis[14] that the unreconstructed Pt(111) surface lies almost at the stability boundary, i.e., it is only just stable. Thus, slight perturbations – as provided by the supersaturated Pt vapour, or by the thermal expansion of interlayer spacings upon heating the crystal – suffice to drive the surface into a reconstructed state. Given the importance of Pt(111) as a catalyst, it is certainly interesting to note this structural instability.

When the chemical potential is lowered below its threshold value of 36 mRy, the surface reconstructs into a pattern characterized by a misfit dislocation domain wall network that forms either a honeycomb pattern or a tiling of wavy triangles. Simulated STM images of these structures bear a striking resemblance to the experimental images. In particular, we too obtain "bright rotors" that result from a slight rotation of the top layer relative to the substrate.

Though it is energetically unfavourable for Pt(111), if the density of the top layer is further increased beyond that at which one sees the wavy triangles, one obtains a domain pattern of "bright stars" and, ultimately, a Moiré pattern. Though these patterns are not seen on Pt(111), they have been seen on various heteroepitaxial surface systems. We see smooth transitions between all these domain patterns, obtaining a unified description for all the different patterns seen on various surfaces.

We believe that the techniques used in this paper can be successfully applied towards designing systems of misfit dislocation patterns with desired spacings, that can be used as templates for growing nanostructures.

Acknowledgments

We thank Stefano de Gironcoli for providing the Pt ultrasoft pseudopotential, and Xavier Gonze and Umesh Waghmare for helpful discussions.

References

1. A.R. Sandy, S.G.J. Mochrie, D.M. Zehner, G. Grübel, K.G. Huang and D. Gibbs, Phys. Rev. Lett. **68** 2192 (1993).
2. M. Bott, M. Hohage, T. Michely and G. Comsa Phys. Rev. Lett. **70** 1489 (1993).

3. M. Hohage, T. Michely and G. Comsa, Surf. Sci. **337** 249 (1995).
4. D.D. Chambliss, R. J. Wilson and S. Chiang, Phys. Rev. Lett. **66** 1721 (1991).
5. H. Brune, M. Giovannini, K. Bromann and K. Kern, Nature **394**, 451 (1998).
6. J. Frenkel and T. Kontorova, Z. Sowjetunion **13**, 1 (1938).
7. S. Baroni, A. Dal Corso, S. de Gironcoli and P. Giannozzi, http://www.pwscf.org
8. D.M. Ceperley and B.J. Alder, Phys. Rev. Lett. **45**, 566 (1980).
9. J.V. Barth, H. Brune, G. Ertl and R.J. Behm, Phys. Rev. B **42**, 9307 (1990).
10. S. Narasimhan and D. Vanderbilt, Phys. Rev. Lett. **69**, 1564 (1992).
11. J.V. Barth, R.J. Behm and G. Ertl, Surf. Sci. Lett. **302** L319 (1994).
12. C. Günther, J. Vrijmoeth, R.Q. Hwang and R. J. Behm, Phys. Rev. Lett. **74**, 754 (1995).
13. J.A. Meyer, P. Schmid and R.J. Behm, Phys. Rev.Lett. **74**, 3864 (1995).
14. S. Narasimhan and R. Pushpa, submitted to Phys. Rev. Lett., 2002.

METALLIC SURFACES UNDER ELEVATED GAS PRESSURE STUDIED IN SITU BY SCANNING TUNNELING MICROSCOPY : O_2, H_2/Au(111); CO/Au(110)

FRANCISCO JOSE CADETE SANTOS AIRES, CYRILE DERANLOT, YVETTE JUGNET, LAURENT PICCOLO, JEAN-CLAUDE BERTOLINI

Institut de Recherches sur la Catalyse (UPR 5401 – CNRS)
2, Avenue Albert Einstein. 69626 – Villeurbanne cedex. France.
E-mail : aires@catalyse.univ-lyon1.fr

We have built an experimental set-up including a scanning tunneling microscope in order to study the behaviour of catalytic surfaces in environmental conditions close to that needed in catalysis (pressure of gas, temperature). We have studied gold surfaces under gas pressure since gold was recently found to be highly active in the low temperature catalytic oxidation of carbon monoxide without combustion of hydrogen. We have followed real-time morphological modifications of Au(111) surfaces under elevated pressures (1 bar) of hydrogen and oxygen. We have also studied the adsorption of CO on Au(110)-(1x2) reconstructed surface and observed in-situ by STM the dramatic change of the morphology of the terraces under the influence of the CO pressure (up to 670 mbar) on the structure of the surface. Complementary XPS and IRRAS measurements were also performed and show CO dissociation the surface.

1. Introduction

Surface structure and catalytic properties of metals are often intimately related. Well-known surface structures (or reconstructions) in UHV conditions may vary and evolve under gas pressure. Scanning tunnelling microscopy (STM) may yield information on surface morphology and structure down to atomic level. It is thus a well suited technique to follow in situ surface modifications due to gas environments at relatively high pressures [1, 2]. In this paper we present results obtained with an STM experimental set-up built in our laboratory [3] in order to study the behaviour of catalytic surfaces in environmental conditions close to that needed in catalysis (pressure of gas, temperature). We have chosen to perform our studies on gold surfaces. We present evidence of surface modifications followed by STM for the systems O_2/Au(111)-glass, H_2/Au(111)-glass and CO/Au(110).

Gold has been generally considered not to be a reactive metal, and therefore a priori not interesting for applications in heterogeneous catalysis. However, recent works have demonstrated the ability of gold to be highly active in the low temperature catalytic oxidation of carbon monoxide without combustion of hydrogen [4] which is a reaction of interest to eliminate traces of carbon monoxide in hydrogen, necessary for applications such as in fuel cells. The active sites to couple favourably the CO molecules, and to dissociate oxygen, are not well understood. It is frequently postulated that the support has to play an important role in this catalytic process. However, it is sometimes proposed that gold itself can be reactive, if it is either in the shape of very small particles or in the shape of surfaces exhibiting sites of particular geometries, with low coordinated atoms. Our work follows this latter hypothesis.

2. Experimental methods

We have built an STM experimental set-up based on an OMICRON MicroLH STM and allowing in-situ observations under gas pressure (10^{-9} - 10^{3} mbar) and variable temperature (120 - 500 K). The set-up consists in two chambers communicating through a gate valve permitting sample and tip transfers with a magnetic drive. The preparation chamber is equipped with an ion gun and a infra-red lamp in order to perform surface cleaning and restructuring (ion bombardment/annealing cycles) as well as high temperature treatments (up to 1273 K). The STM chamber contains the Micro LH and the sample/tip carrousel. The Micro LH STM has been slightly modified in order to be compatible with pure gas high pressure (O_2, H_2, ...) : gold plating of the parts sensitive to oxidation and/or hydrogenation (heating wires, magnets) and protection of the piezoelectric tube. The latter allows scans up to 15 μm and height measurements up to 2 μm. Electrochemically etched PtIr (80:20) tips were used for the STM observations. The vibration damping is assured by four air "cushions". The gas introduction line is equipped with filters for impurities (H_2O, ...) present in the commercial gases. In the case of CO particular care was taken in order to avoid as much as possible nickel deposition on the surface through decomposition of nickel carbonyls either present in the CO sources or produced in the gas line.

Concerning the Au(111)-glass, sample preparation is rather simple. The as-deposited Au film on glass is rough; in order to obtain a well organized surface with flat terraces (Figure 1) ex-situ flame-annealing in air or in situ annealing under UHV have been performed with similar results.

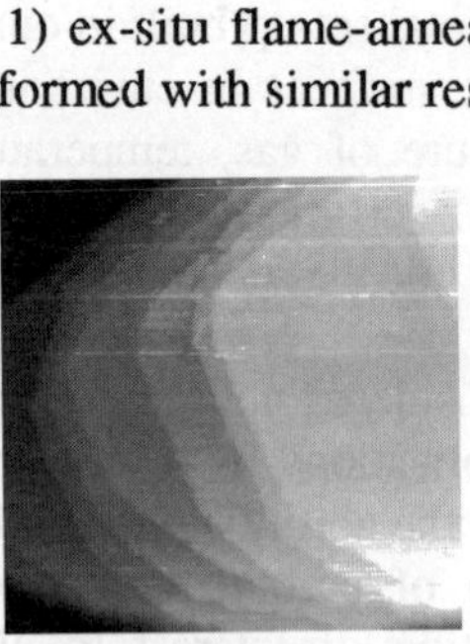

Figure 1. Large scale (200 nm x 200 nm) STM topographic image of the [111]-oriented Au surface. All steps are monoatomic (0.24 nm)

The Au(110) single crystal was cleaned in UHV by repeated series of argon ion sputtering and annealing at 610K which is lower than the 3D-roughening transition temperature and yields a (1x2) reconstruction [5].

During STM experiments on both surfaces, variation of tunnelling current (0.1 – 2 nA) and of bias (0.02 – 1V) did not induce modifications on the images.

In the case of CO adsorption on Au(110) reflection absorption infrared spectroscopy (RAIRS) was also performed in a separate set-up. RAIRS spectra were obtained with a Thermo Nicolet NEXUS Fourier transform infrared spectrometer (FTIR) in specular reflection mode under grazing (8°) incidence. A ZnSe polarizer is inserted in the path of the incoming beam in order to select either the s-polarized (perpendicular to the plane of incidence) or the p-polarized (parallel to the plane of incidence) component of the light. Acquisition of spectra with p and s polarization (Ip and Is) followed by calculation of the ΔR/R= (Ip-Is)/(Ip+Is) yields direct information on the adsorbed species since the contribution of the gas phase (isotropic and insensitive to polarization) is eliminated from the spectra. The surface was controlled by XPS and/or AES before and after the adsorption experiments.

3. Results and discussion

3.1 STM performances under UHV and high pressure conditions : O_2, H_2/Au(111)

Under high vacuum conditions (2 10^{-9} mbar), atomic resolution has been readily obtained on the [111]-oriented terraces (Figure 2a). Distances around 0.29 nm corresponding to <110> directions are observed. The corrugation is less than 0.02 nm. At lower resolution the "herringbone" reconstruction characteristic of the Au(111) surface [4] could also be observed (Figure 2b). Corrugation is 0.03 nm.

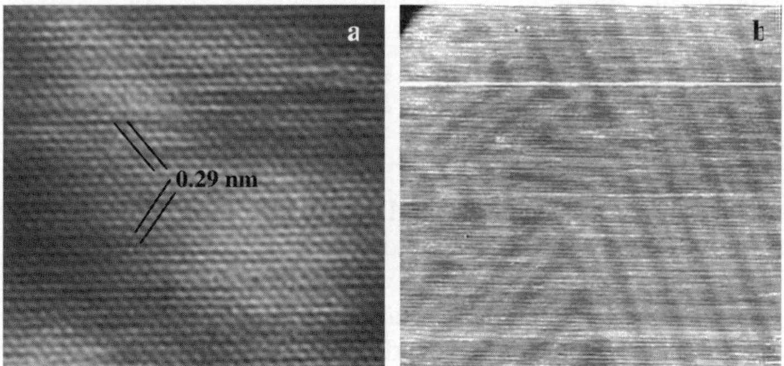

Figure 2. (a) atomic resolution STM image (8 nm x 8 nm) obtained on Au(111) terraces. (b) the "herringbone" reconstruction on the Au(111) surface (image : 75 nm x75 nm)

High resolution STM imaging of the Au(111) surface exposed to high pressures (1 bar) of oxygen and hydrogen becomes more difficult than under UHV. This is mostly related to the fact that atoms and islands on the surface are rather mobile. Nevertheless, we could achieve atomic resolution under 1bar of O_2 at 370K (Figure 3a) and we were able to observe the "herringbone" reconstruction under 1 bar of H_2 at 330K (Figure 3b). When compared to UHV results, these observations do not show any peculiar structural differences.

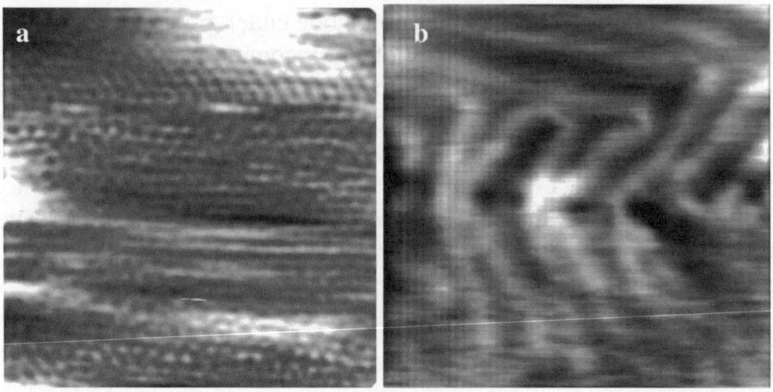

Figure 3. (a) Topographic STM image (8 nm x 8 nm) showing atomic resolution in a Au(111) surface under oxygen pressure (1 bar) at 370K. (b) "Herringbone" reconstruction of the Au(111) surface under 1 bar of hydrogen at 330 K (image is 33 nm x 33 nm).

However, in the presence of oxygen and hydrogen we have observed dynamic effects on the surface. In Figure 4, obtained at room temperature under 1 bar of oxygen, we present 6 consecutive images (the original film contains 93 images) from a region where we were able to follow in real-time (one image every 8 seconds) the diffusion of monoatomic-high islands and their re-arrangement with the terraces. Similarly, under 1 bar of hydrogen at room temperature, it was possible to follow the coalescence of two monoatomic-deep holes on a terrace (Figure 5) which is possible if the atoms at the borders of the "hole" are mobile. This clearly shows the mobility of the gold surface under gas pressure.

Concerning the performances of our system we have seen that structural information can be obtained from UHV to 1 bar under controlled environment and that morphological changes due to a given environment can be followed in situ in real-time (one image in several seconds). This allows molecular adsorption studies on metallic surfaces at high pressures which are more compatible with catalytic reaction conditions than UHV.

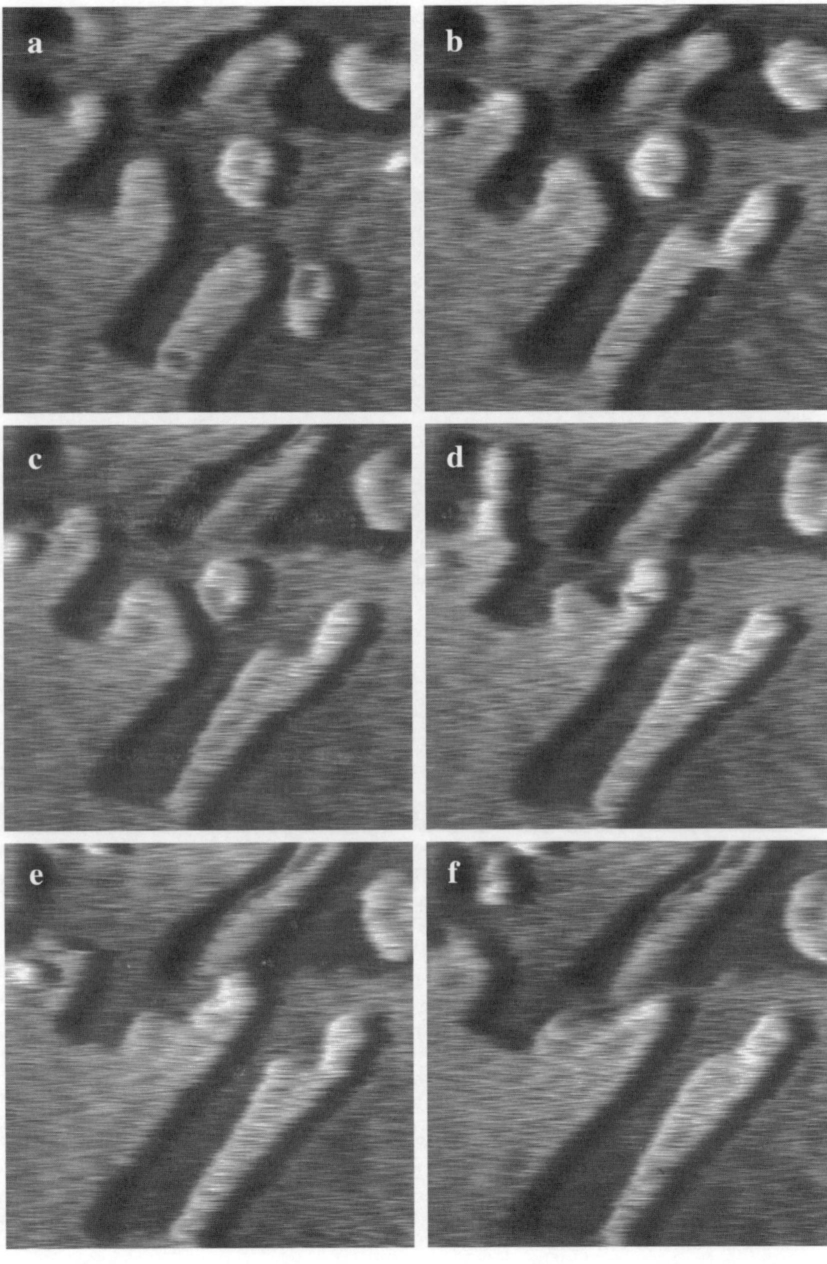

Figure 4. (a-f) Evolution of the gold surface under 1 bar of oxygen at room temperature. (51 nm x 51 nm images acquired every 8 seconds). Height of steps and islands is monoatomic (0.24 nm).

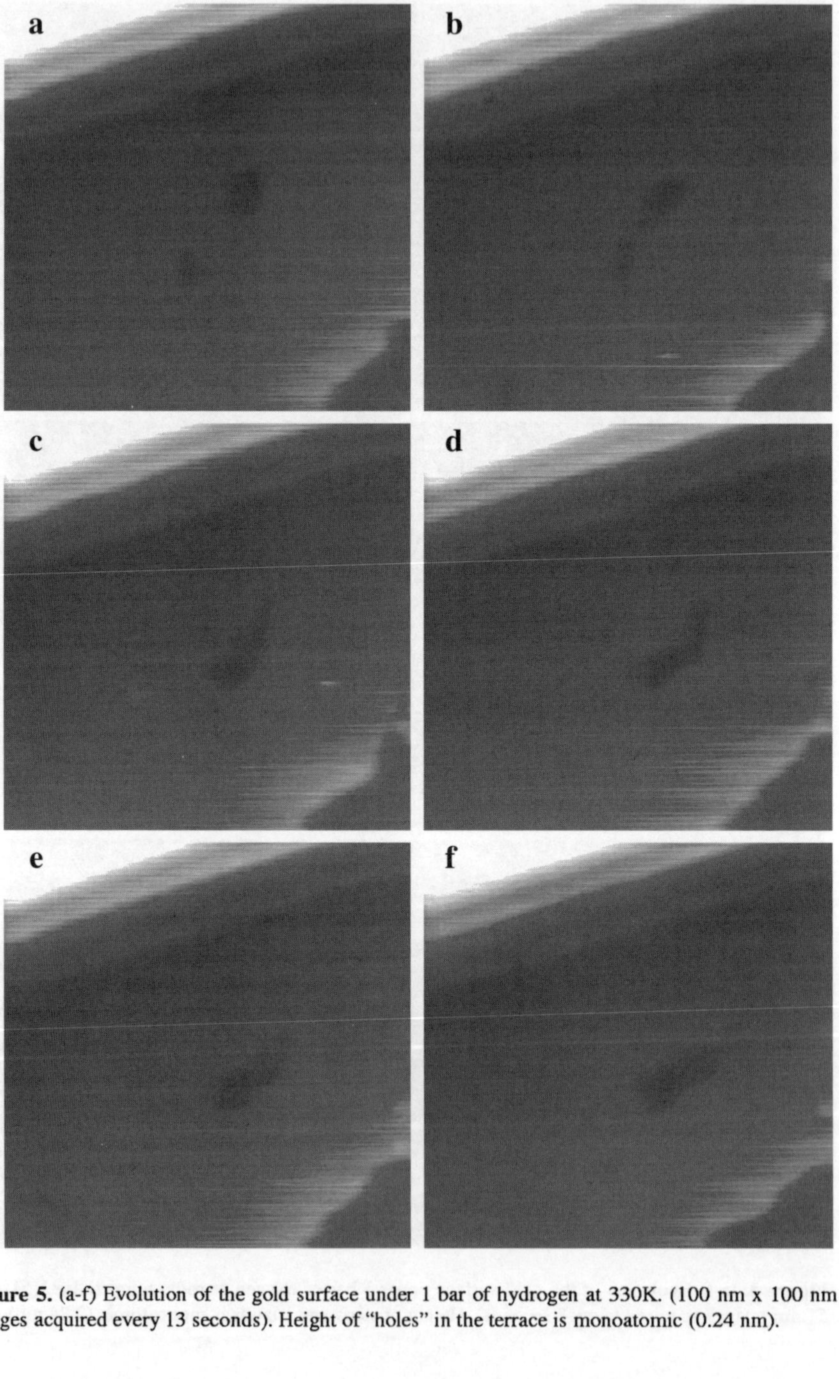

Figure 5. (a-f) Evolution of the gold surface under 1 bar of hydrogen at 330K. (100 nm x 100 nm images acquired every 13 seconds). Height of "holes" in the terrace is monoatomic (0.24 nm).

3.2 CO/Au(110)

Au(110) surface reconstructs in a (1x2) missing-row geometry when it is annealed under vacuum at temperatures around 610K (Figure 6a-insert). The morphology of the surface is presented in Figure 6a where large anisotropic terraces (height : 0.15 ± 0.02 nm) aligned in the <1 $\bar{1}$ 0> can be observed. This alignment is directly related to the (1x2) reconstruction. After exposure to CO and, whatever the pressure (1 – 670 mbar), this anisotropy is lifted and the monoatomic-high terraces are smaller and have a more isotropic shape (Figure 6b-c). This indicates a deep restructuring of the surface induced by the adsorption of CO. Equivalent pressures of nitrogen do not induce restructuring of the surface. The height of the terraces observed after restructuring is identical to that observed before restructuring (0.15 ± 0.02 nm) and remains consistent with the distance between two successive Au(110) planes. This restructuring occurs very rapidly (within less than 5 minutes under CO).

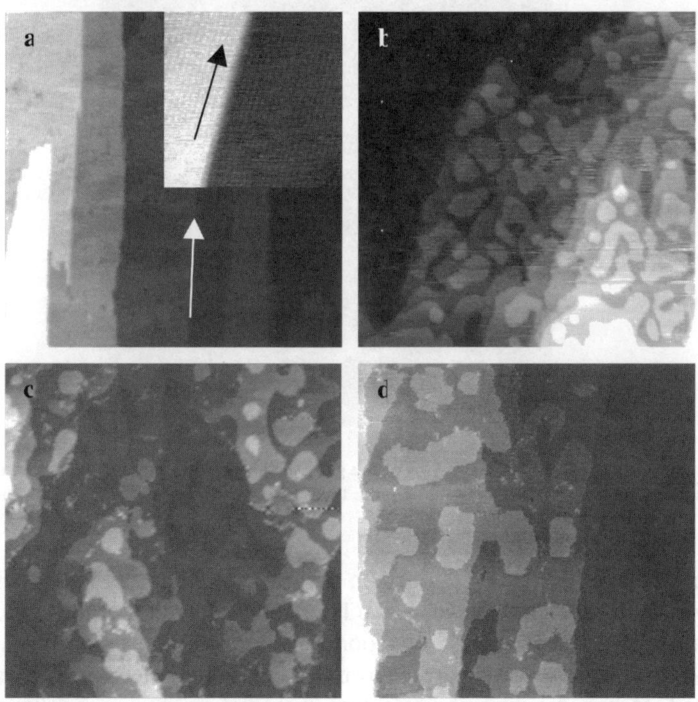

Figure 6. Evolution of the gold surface with increasing CO pressure. (a) 10^{-9} mbar, insert : high resolution image showing the (1x2) reconstruction with 0.82 nm distances between <1 $\bar{1}$ 0> rows, arrows indicate the <1 $\bar{1}$ 0> direction ; (b) 1 mbar; (c) 40 mbar, (d) 670 mbar. Images are 350 nm x 350 nm (except insert in (a)). Height of steps and islands is monoatomic (0.15 nm).

We believe that lifting of the anisotropy of the terraces can be reasonably associated to a transition from a (1x2) to a (1x1) surface atomic arrangement although this has not been directly observed in the STM images. After evacuation to 3 10^{-9} mbar, the surface morphology remains unchanged, showing the previously described isotropic terraces (Figure 7). However, a roughening of the surface of the terraces is now observed in the STM images. The roughness details are randomly distributed on the terraces and have heights around 0.05 nm. In parallel to these observations RAIRS experiments performed in the same pressure conditions show the presence of linearly adsorbed CO. The intensity of the CO stretching contribution decreases with time of exposure to CO until it completely disappears. Following this, further adsorption of CO is no longer possible. After these experiments XPS and AES analysis of the surface show the presence of carbonaceous residues that can be associated with the roughness details observed in the STM images (Figure 7) [6].

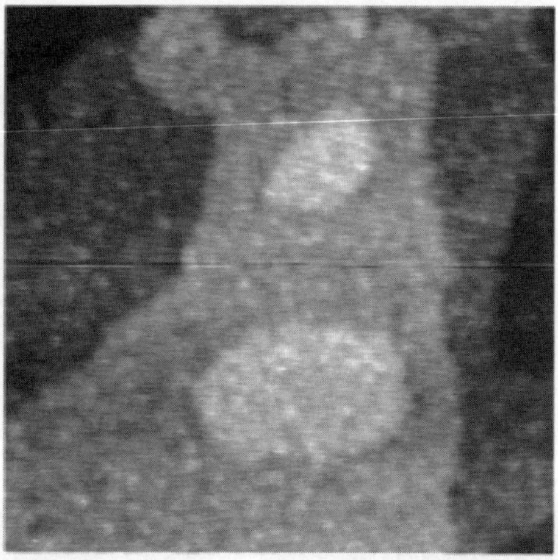

Figure 7. STM image (150 nm x 150 nm) of the gold surface after exposure to CO and evacuation 3 10^{-9} Height of steps and islands is monoatomic (0.15 nm). Roughness details are around 0.05 nm high.

The presence of carbon residues on the surface indicates the dissociation of CO on Au(110). Such behaviour is can be associated with fact that the Au(110) surface is an open surface with low-coordinated atoms that can be compared to kink and step atoms which are known to be more reactive. The randomly distributed carbon residues indicate that dissociation occurs on the terraces. Dissociation most probably proceeds via the Boudouard reaction following 2CO $\rightarrow CO_{2,g} + C_a$. A similar behaviour has been recently observed for Pt single crystals [7].

Conclusion

Our STM experimental set-up allows structural and morphologic studies of surfaces under high pressures (up to 1 bar) of gases opening possibilities to study adsorption and reaction in conditions close to that of catalytic reactions.

CO adsorption on Au(110) at pressures between 1 and 670 mbar induces a deep restructuring of the surface observed by STM. Equivalent RAIRS experiments show a decrease with time of CO stretching contribution leading to the presence of carbon residues on the surface detected by XPS and imaged by STM. These results lead to believe that CO dissociation occurs on Au(110) surface. This peculiar reactivity could be a specificity of open (110) faces which could have important issues in catalysis by gold.

References

1. J.A. Jensen, K.B. Rider, M. Salmeron and G.A. Somorjai, *Phys. Rev. Lett.* **80**, 1228 (1998).
2. L. Osterlund, P.B. Rasmussen, P. Thostrup, E. Laegsgaard, I. Stensgaard and F. Besenbacher, *Phys. Rev. Lett.* **86**, 460 (2001).
3. F.J. Cadete Santos Aires, C. Deranlot, in *Proceedings of the 12th European Congress on Electron Microscopy*, vol. III Instrumentation and Technology, Brno, july 2000, edited by P. Ciampor, L. Frank, P. Tomanek, and R. Kolarik, pp. 1263.
4. M. Haruta, N. Yamade, T. Kobayashi and S. Ijima, *J. Catal.* **301**, 299 (1989).
5. M. Sturmat, R. Koch, and K.H. Rieder, *Phys. Rev. Lett.* **77**, 5071 (1996).
6. Y. Jugnet, F.J. Cadete Santos Aires, C. Deranlot, L. Piccolo and J.C. Bertolini, *Surf. Sci. Lett.* **521**, L639 (2002).
7. K. McCrea, J.S. Parker, P. Chen and G. Somorjai, *Surf. Sci.* **494**, 238 (2001).

X-RAY STRUCTURAL ANALYSIS OF SEMICONDUCTOR-ELECTROLYTE INTERFACES

S. WARREN, T-L. LEE AND J. ZEGENHAGEN
ESRF
Avenue Jules Horowitz
BP-220 F-38043 Grenoble FRANCE

A. REITZLE, D.M. KOLB AND J. ZIEGLER
Abteilung Elektrochemie,
Universität Ulm
D-89069 Ulm, GERMANY

F. MAROUN AND P. ALLONGUE
CNRS UPR 15, Physique des Liquides et Electrochimie
Université Pierre et Marie Curie
4 Place Jussieu, 75005 Paris, France

A. KAZIMIROV
CHESS, Wilson Laboratory
Cornell University, Ithaca
New York 14853, USA

G. SCHERB
Max-Planck-Institut-FKF
Heisenbergstrasse 1
D-70569 Stuttgart, GERMANY

The *in-situ* and *ex-situ use of* x-ray methods to investigate the structure of semiconductor-electrolyte interfaces is discussed, with examples of several systems studied. The design of electrochemical cells for investigating *in-situ* potential controlled and dynamic electrochemical processes is described, together with applications of their use. The variety of metal structures availble from electrochemical depostion is shown, by describing Cu deposition on GaAs, electroless deposition of Au on Si, and potential controlled deposition of Au on Si. Examples of the benefit of using surface x-ray diffraction, and x-ray standing waves to investigate metal depostion on semiconductor surfaces are given.

1. Introduction

Metal deposition on semiconductors is of great interest due to the importance of ohmic and Schottky-type contacts in microelectronics. Considerable progress has been made in the understanding of evaporation procedures for metal deposition by surface and structural studies under UHV conditions. In contrast, electrochemical deposition of metals onto semiconductor surfaces has been less widely studied, yet this represents a low cost “all-wet” preparation route that may be easily integrated into the chemical processing of semiconductor surfaces. Monitoring of the electrochemical characteristics (e.g. potential - current dependencies) [1] of the system during deposition does not give direct information about the *structure* of the surface-electrolyte interface and the deposit. Therefore, studies of semiconductor surfaces during electrodeposition by structure sensitive techniques, especially *in-situ*, are vital. Metal electrodes have been studied extensively by both *in-situ* and *ex-situ* techniques [2,3] and this has led to a great deal of information regarding the mechanism of electrochemical reactions at metal surfaces. A similar approach to semiconductor electrode surfaces is hoped to bring a greater understanding of the processes occurring during metal deposition.

The fundamental aspects of metal growth by electrochemical deposition are also of great interest, as the applied electrode potential, and electrolyte composition can be used to influence the nucleation and growth mode. The distinctive feature of electrochemical control is that it allows the electrode potential, and therefore the electrochemical potential at the substrate surface, to be carefully controlled, which allows metal deposition both at, and far away from, equilibrium conditions. The examples below describe a wide range of metal deposit structures, including isolated atoms in ordered ‘wetting layers’ or substituted in the semiconductor atom sites, and epitaxial islands, or clusters aligned along stepped surfaces.

Surface structure sensitive techniques such as atomic force microscopy (AFM) and scanning tunnelling microscopy (STM) may be used to retrieve useful information regarding the shape and distribution of the metal deposit. However, *in-situ* STM for semiconductor-electrolyte interfaces is often very difficult. Furthermore, surface x-ray diffraction (SXRD) and crystal truncation rod (CTR) scattering gives information not accessible by other experimental methods, for example, the ‘roughness’ of the substrate, and the lattice structure and correlation lengths of deposited clusters. A detailed review of the use of SXRD and CTR measurements in sample characterization has been given in [4]. X-ray standing waves (XSW) measurements produce element specific information regarding the position, and degree of ordering of the deposited layer. The theory and applications of such experiments has been described in [5].

2. Experimental

2.1. *Characterization of samples*

In addition to the electrochemical information recorded e.g. during cyclic voltammetry, AFM imaging was used for structural information, and to determine the shape and distribution of deposited islands. All *ex-situ* atomic force microscopy (AFM) experiments (unless otherwise stated) were performed in air with a Pico-SPM (Molecular Imaging Corporation, Tempe, USA) that operates in contact mode, using Si cantilevers (Park Ultralever 0.6 μm) with a spring constant of 0.31 Nm^{-1} as specified by the manufacturer. Rutherford Back Scattering (RBS) was used to obtain information about the amount of metal deposited, and was carried out after sample measurement, to calibrate the coverage of each series of samples.

2.2. *Description of beamlines*

SXRD data were recorded using the z-axis diffractometer at the BW2 wiggler beamline HASYLAB, Germany as described previously [6]. The XSW experiments were carried out at beamline X15A of the National Synchrotron Light Source, Brookhaven National Laboratory, USA, and the experimental set-up is described in [5]. More recent SXRD and XSW measurements were carried out at beamline ID32, ESRF, Grenoble. ID32 is a windowless beamline, with 2 undulators, operating over a photon energy range of 2.5 to 40 keV, equipped with a Si(111) monochromator, and a plane mirror for vertical focusing of the beam and rejection of higher harmonics from the double-crystal Bragg-monochromator. The photon flux at the sample position is 10^{13} photons/sec at 10 keV photon energy and an electron beam current of 200 mA with an unfocussed beam size of 400x800μm (horizontal x vertical). For SXRD and XSW in air, the beamline is equipped with a four-circle diffractometer, a scintillation X-ray point detector, and a Si drift diode fluorescence detector.

2.3. *Electrochemical cells for in-situ x-ray studies*

At all beamlines, *in-situ* sample measurements were carried out using one of the electrochemical cells shown in Figure 1, which may be easily mounted and aligned on the diffractometer. A 'thin-layer geometry' cell with a Mylar foil window was used to study various semiconductor-electrolyte systems. The design and operation of this cell has been described previously [7] and essentially allows the study of electrochemical systems without mass transport at different applied electrode potentials. To allow mass transport and diffusion, the

Mylar window is inflated, and then deflated when the X-ray measurements are carried out, whilst maintaining potential control. The thickness of the electrolyte trapped between the window and the sample is of the order of 1-10 μm. A 'thick layer' in-situ cell with bulk electrolyte above the sample allows the study of dynamical electrochemical processes [8]. A hemispherical quartz dome (thickness ~ 100μm), filled with electrolyte, surrounds the sample. The enhanced absorption owing to the quartz and the 8 mm path through the electrolyte requires measurements to be carried out at photon energies above 15 keV.

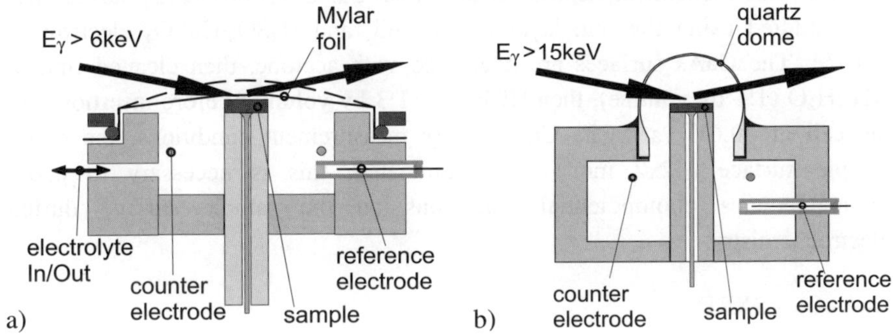

Figure 1 in-situ electrochemical cells with (a) thin layer and (b) thick layer geometry

3. Results and Discussion

3.1. *GaAs*

GaAs stripping

The preparation of smooth GaAs surfaces in UHV using 'capped' samples has been well studied and characterised [9-11]. A MBE grown GaAs film is capped with an amorphous As layer, which allows exposure to ambient atmospheres, and the As layer can later be removed by annealing in UHV. A comparable preparation route using electrochemical methods to strip the As capping layer was devised.

Using the thick layer geometry *in-situ* cell [7] and a high purity 0.5M H_2SO_4 electrolyte, the reflectivity of the surface GaAs (220) peak, i.e. the (220) crystal truncation rod, was monitored at a small angle of incidence and at a constant potential of –1.2V vs SCE [12]. An increase of the reflectivity occurs over a period of around 2 hours, which is explained by the gradual stripping of the amorphous As capping layer. A plateau in the reflectivity, which occurs simultaneously with an increase in the electrochemical current is assigned to the stripping of one or two layers of ordered As. The first few As layers on the GaAs surface are well ordered, and therefore more difficult to strip than the amorphous 'bulk' As material. Thus, the well defined, MBE surface is recovered at the onset of the plateau. The thick layer geometry is required in these measurements, as the stripping process occurs via electrochemical generation of atomic hydrogen, which then etches the capping layer, producing gaseous AsH_3. The large volume of electrolyte above the sample surface is essential since it permits diffusion in and out of both reaction partners and products.

GaAs:Cu

The electrochemical deposition of Cu on the GaAs(001) [13] surface has been studied using the thin layer *in-situ* cell, and $H_2SO_4/CuSO_4$ electrolytes [14,15]. The GaAs surfaces are degreased with acetone, then cleaned first in NH_3:H_2O (1:1 by volume), then HCl:H_2O (1:3 by volume) before insertion into the cell at –0.6V vs. SCE. Under x-ray measurement conditions, the entire sample surface (~2x2 mm^2) was illuminated; this is necessary to avoid inhomogeneous photopotential conditions on the sample surface during electrochemistry.

GaAs:Cu - SXRD

After deposition from 1mM $CuSO_4$+0.5M H_2SO_4 at –0.55V vs. SCE, in-plane SXRD scans of the GaAs(001) surface clearly show the presence of 3D Cu crystallites. These grow with Cu[100] aligned along the GaAs[110] direction, and Cu[110] aligned with GaAs[100] [12,14]. This represents an unstrained fcc Cu lattice epitaxial with the GaAs(001) surface. The mosaic spread of the crystallites in-plane is around FWHM $\Delta\omega$= 4.5°, indicating a significant deviation from perfect alignment. The lateral size of the clusters is estimated at around 30nm, and the vertical height around 10nm, (or around 55 layers of Cu in the fcc packing geometry) as measured from the $Cu(111)_{cubic}$ Bragg reflection. The total coverage deduced from the electric charge measured during deposition was 50 dense packed Cu layers, i.e. 50 monolayers of $1.5x10^{15}$ atoms cm^{-2} each on the GaAs(001) surface. The fact that the islands are just slightly thicker means they have almost coalesced. The width from transverse scans through the $Cu(111)_{cubic}$ Bragg reflection is around 20°, and 2 maxima are seen in the ω scans. More careful investigation of the alignment of the Cu crystallites was performed by scans at constant q (scanning motors μ and ω) showing a titled geometry of the crystallites on the GaAs(001) surface, with the tilt angle around 6°. No azimuths of the plane seem to be preferred, with crystallites tilted around the epitaxial direction in a homogeneous ring as shown in Figure 2.

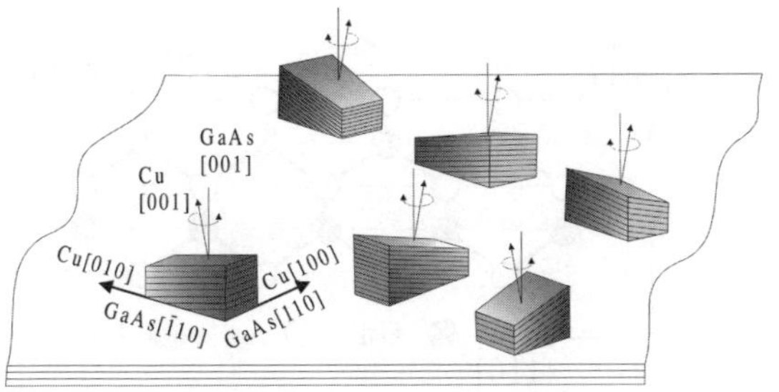

Figure 2 Schematic showing the geometry of the Cu clusters on the GaAs(001) surface, aligned along the GaAs[110] direction, and tilted with respect to the substrate surface normal

GaAs:Cu - XSW

The first (sub)monolayers of Cu deposited electrochemically on the GaAs(001) surface were found to occupy different sites from the majority of the deposit formed at higher coverages (when cluster formation is preferred). There is some diffusion of the Cu atoms a few nm into the bulk of the GaAs substrate, which is more significant for n-type than for p-type samples. For an investigation of atomic positions in the submonolayer range, XSW has been used to identify the adsorption sites and occupancies [16]. Both n-type and p-type GaAs substrates have been used, which were cleaned and then mounted in the thin-layer electrochemical *in-situ* cell, using the same electrolyte as for the SXRD measurements.

The coherent position values P measured for the (004) reflection normal to the surface suggested a substitutional Ga site for Cu [16]. This is confirmed by the off-normal XSW measurements, shown in Figure 3. The coherent fraction F for both n-type and p-type substrates and for both reflections was relatively low ($F \leq 0.65$), indicating that although the substitutional site is preferred, some Cu atoms also occupy other sites within the GaAs host lattice. The coherent fraction F decreases with the amount of Cu on the sample, and the coherent position P changes slightly (the lowest coverage measured (0.2ML) showed P= 0.98, and above 1ML P=0.90) indicating that the diffusion and substitution processes are coverage dependent, with minority sites occupied at higher coverages. A diagram of the GaAs lattice, indicating schematically the diffraction planes utilized for the XSW measurements and the sites, occupied by Cu, is shown below in Figure 3.

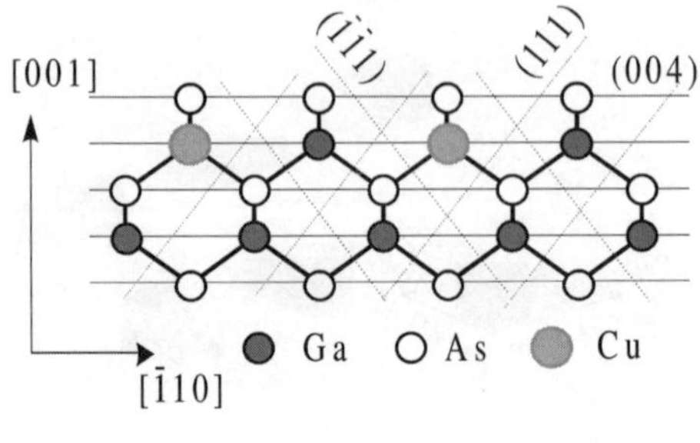

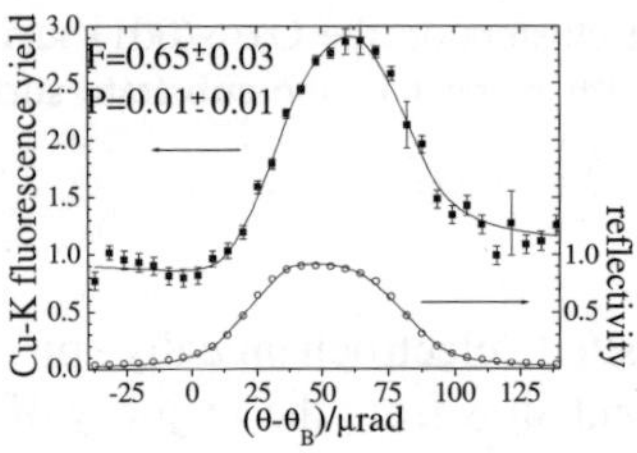

a)

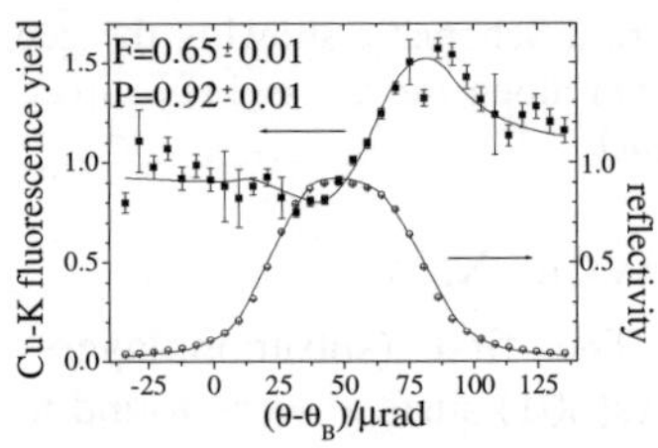

b)

Figure 3 XSW scans recorded at 10.2KeV for Cu deposited on GaAs(001) (a) 0.5 ML Cu coverage for the GaAs(111) reflection, (b) 1 ML Cu for the GaAs(-1-11) reflection. In both cases, P=0 corresponds to the Ga sites. (c) Shows schematically the orientation of the lattice planes and the Cu atoms occupying Ga sites in the GaAs lattice.

c)

GaAs:Cu- Discussion and Summary

The initial deposition of Cu proceeds by diffusion and substitution of Cu onto the Ga lattice site. As the coverage of Cu increases, clusters are formed which are aligned in-plane to the GaAs surface high symmetry directions, but which grow tilted with respect to the substrate surface normal.

3.2. *Hydrogen terminated Si(111)*

Flat monohydride-terminated Si(111) surfaces are easily produced from clean, oxide free Si substrates, by chemical etching in NH_4F [17]. The quality of these surfaces can be examined using AFM and SXRD to investigate surface morphology. Although Si is cubic, the Si(111) surface can be more conveniently represented by a hexagonal unit cell, so that the directions H and K are lying within the (111) surface plane, and L represents the out-of-plane scattering direction. In this case, the reciprocal space unit cell used is: $a_{hex}^* = 1.8892$ Å^{-1}, $b_{hex}^*=1.8892$ Å^{-1}, $c_{hex}^*=2.004$ Å^{-1}, $\alpha=\beta=90°$ and $\gamma=60°$. The hexagonal vectors are related to the cubic vectors by the following transformations, where $|a^*_{cubic}| = 2\pi/a_{Si}$, and $a_{Si} = 5.431$ nm:

$$a^*_{hex} = \frac{2}{3}\begin{pmatrix} 1 \\ \frac{1}{2} \end{pmatrix} |a^*_{cubic}| \quad b^*_{hex} = \frac{2}{3}\begin{pmatrix} \bar{1} \\ 2 \\ 1 \end{pmatrix} |b^*_{cubic}| \quad c^*_{hex} = \begin{pmatrix} 1 \\ 1 \\ 1 \end{pmatrix} |c^*_{cubic}|$$

and thus: $$d_{cubic} = \frac{\sqrt{3}\, a_{Si}}{\sqrt{8(h^2 + k^2 + hk) + 9l^2}}$$

relates the real space cubic lattice spacing to the hexagonal h, k, l values. Hexagonal coordinates will be marked in the following by the subscript *hex*.

The measurement of the crystal truncation rods (CTR) from the Si(111) surface is therefore carried out by an l scan across the two Bragg peaks at l=1/3 and l=4/3, as shown in Figure 4. In the region between the Bragg-reflections, the scattered intensity is strongly dependent on the surface roughness. By fitting the experimental data to the model by Robinson [18], one can derive a β value, which describes the occupancy β^n of the surface level n. From this β value, a root-mean-square value of the surface roughness σ_{RMS} can be calculated:

$$\sigma_{RMS} = \frac{\sqrt{\beta}}{1-\beta}\, d_\perp$$

where $d_\perp$ (3.14 Å) is the distance between the Si(111) planes. For this surface we obtain a surface roughness value σ_{RMS} of 4.2 Å in air, and almost the same value (4.31 Å) from the hydrogen terminated Si(111) surface in electrolyte solution, indicating that any damage to the hydrogen termination caused by the electrolyte is localized, and does not detrimentally affect the surface structure.

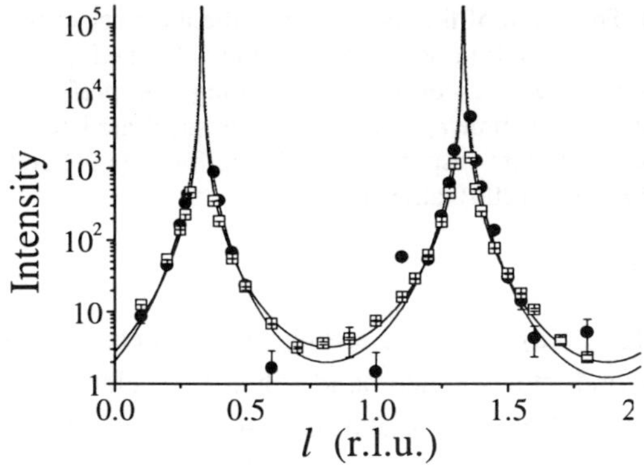

Figure 4 CTR from Si(111):H surface in air (□) and in contact with electrolyte (●). The solid line and dotted line show the fit to the experimental data in air and in electrolyte respectively. The data from the electrode in solution have been scaled to account for the absorption of x-ray intensity by the electrolyte.

Pb-H-Si(111)

Lead deposited onto hydrogen-passivated Si(111) surfaces is an excellent model system, as no interdiffusion or silicide formation is observed. *Ex-situ* AFM studies indicated a 3D growth mode of Pb clusters, and *in-situ* XSW confirmed an entirely 3D Volmer-Weber nucleation and growth process [19]. The structure and epitaxy of the deposited Pb has been studied, and also the dynamics of the Pb cluster growth process, thus both the thin layer, and thick layer electrochemical cells have been used for *in-situ* x-ray diffraction and XSW [19]. Electrolytes were prepared from Milli-Q water, and ultrapure chemicals, and consisted of 0.1M $HClO_4$+ 1mM or 0.1mM $Pb(ClO_4)_2$. The Pb clusters grow predominantly epitaxially aligned with the Si(111) planes, with a minority of clusters with Pb(100) and Pb(110) planes parallel to Si(111). There is a strong preference for in-plane alignment along the high symmetry axes of the Si surface, as shown in figure 5, though this orientation is disturbed when very rough Si(111):H surfaces are used. Studies of the dynamics of cluster growth indicate a ripening process, whereby large Pb islands increase in size, at the expense of the smaller islands.

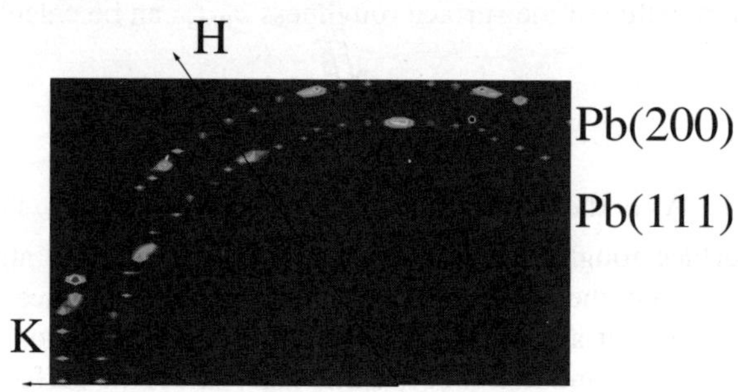

Figure 5 The alignment of the Pb clusters in reciprocal space is shown, relative to the Si(111) hexagonal lattice (H and K). The result of the $\Delta\omega$ scan, i.e., the intensity of the Pb(111) and Pb(200) peaks, which arises from Pb crystallites with the (111) and (200) planes normal to the surface is shown as a color map, illustrating the preferred orientation of the crystallites along the high symmetry directions of the substrate lattice. Maxima of intensity can be identified ±30° with respect to the H-axis for Pb(111) and at ±15° and ±45° with respect to H for the Pb(200) lattice planes.

Electroless Au deposition on Si(111):H

For noble metals in solution (e.g. Au), there is a strong overlap between the metal acceptor state and the semiconductor valence band. Therefore, electroless deposition occurs via hole injection from the metal into the semiconductor valence band. As a consequence, the semiconductor is oxidized as the noble metal is deposited [20, 21]. This electroless deposition method (using an electrolyte of 0.1 M HCl, 5 vol.% HF and 1μM – 1mM K[AuCl4]) was used to deposit Au on hydrogen terminated Si(111) and the net process is:

$$AuCl_4^- \rightarrow Au + 4Cl^- + 3h^+$$
$$Si + 2h^+ \rightarrow Si^{2+}$$

The amount of gold deposited on the surface is expressed in monolayers (ML) where 1 ML = 7.84 $\times 10^{14}$ atoms/cm^2 (the number of silicon substrate surface atoms). AFM images clearly show 3D islands formed on the Si(111):H surface, even at very low Au coverage [22].

Electroless Au deposition on Si(111):H – SXRD

In the specular (out-of-plane) geometry, the Au(111) peak is clearly seen, indicating that (111) planes of gold grow parallel to the Si(111) planes. This is illustrated in Figure 6 for a (nominally) 15.3 ML sample, and Pendellösung fringes are observed, indicating that the island thickness (approximately 23 ML) is well defined, and astonishingly uniform for all the Au clusters on the surface. This suggests a 3D growth mode on the silicon substrate. The in-plane orientation has also been examined, and shows the gold islands are highly aligned to the Si substrate surface, along the high symmetry directions, such as <110>.

The crystal truncation rods of a bare Si(111):H surface and gold-deposited surfaces were also recorded, and an example is shown in Figure 6b. The $(1.34,0,l)_{hex}$ crystal truncation rod (CTR) of the 1.9 ML sample (open triangles) can be compared with the $(1.34,0,l)_{hex}$ CTR of an 'as prepared' H-terminated n-Si(111) surface (open circles) to get information about the roughening of the Si substrate during the gold deposition process. The β values for the Au-free and Au-covered surfaces are β = 0.283 and β = 0.732 respectively, and the corresponding σ_{RMS} increases from 2.3 Å to 10 Å. This is in accordance with the electroless deposition, which proceeds by simultaneous Au deposition and Si etching. The roughening of the surface meant that imaging of samples with high coverages of Au was not possible by AFM.

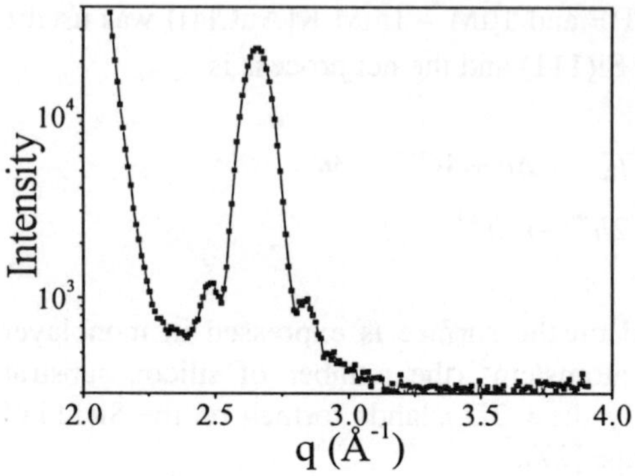

Figure 6
(a) Specular scan along $[00l]_{hex}$ after gold deposition of 15.3 ML, showing the Laue oscillations, indicating well-defined island thickness.

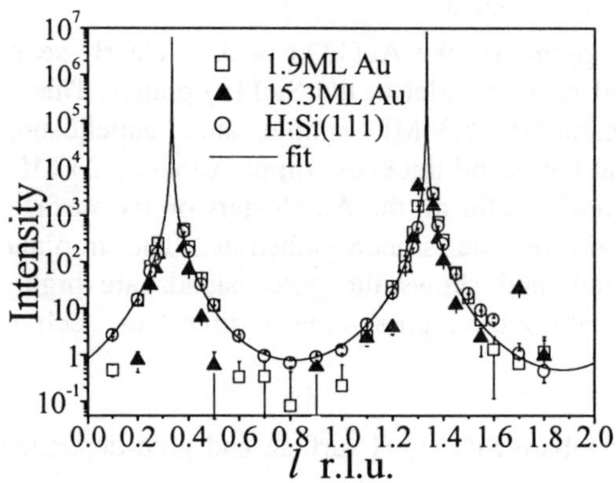

(b) CTR along $[1.34,0,l]_{hex}$ from 1.9 and 15.3 ML Au samples, and the bare H:Si(111) surface. The solid line shows the fit to the data.

Electroless Au deposition on Si(111):H – XSW

The XSW data recorded from a 0.03ML Au sample is shown in Figure 7, with a coherent position P= 0.81 and coherent fraction F= 0.56. The coherent fraction F is significant at these low coverages, and the Au site is found from triangulation with XSW measurements using the Si(022) reflection to be 'atop' the substrate atoms. At increasing coverages the coherent fraction F decreases, becoming very low at higher coverage, indicating a large degree of incoherence with respect to the substrate atomic planes, in agreement with the finding from SXRD that thicker Au clusters are formed [22].

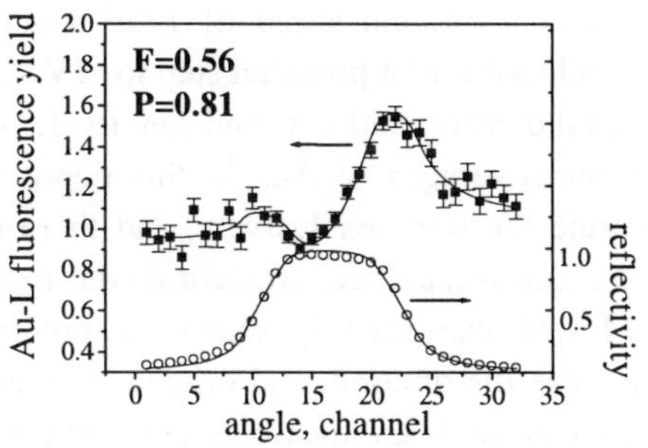

Figure 7 XSW measurements recorded from a 0.03ML Au deposited sample. Reflectivity data (O) and the X-ray standing wave modulation of the Au L x-ray fluorescence (■) are shown, together with the coherent position P (0.81), and coherent fraction F (0.56) obtained from curve fitting to the XSW data, recorded using the Si(111) reflection at 17.2 keV.

Electroless Au deposition – Discussion

The AFM images and the SXRD data indicate cluster formation even at low coverage. The Au islands appear to grow to a uniform, well defined height, with the (111) Au atomic planes parallel to those of the Si(111) substrate, with only a small mosaic spread (FWHM $\Delta\omega$=5.67°). The orientation of the islands within the surface plane is also preferred along the high symmetry $<100>_{hex}$, $<010>_{hex}$, and $<110>_{hex}$ axes. This agrees well with previous work on Au deposition on monohydride Si(111) [23-25], and suggests a 3D or Volmer-Weber growth mode. However, the low coverage XSW data indicate a 2D growth mechanism, with the deposited Au atoms occupying sites 'atop' Si atoms. The Au deposition and growth mechanism therefore seems to be a 2D layer of gold atoms at low coverage, with some clusters already present (as observed in AFM). As the coverage increases, gold clusters are increasingly formed, and the XSW technique can no longer identify the Au position, because the Au atom spacing within the gold island is not coherent with respect to the substrate atomic planes.

Potential controlled Au deposition on miscut Si(111):H

Single height steps (aligned along the $[11\text{-}2]_{cubic}$ direction and separated by terraces of 800-900 Å width) are produced by careful control of the etching conditions and the miscut angle of the Si(111) surface (0.2° along the $[11\text{-}2]_{cubic}$ direction) [26, 27]. The electrode potential during the deposition of Au allows selective deposition at the step edge sites and control of the island density, and is thought to influence the degree of epitaxy and/or specific orientation of the islands. RBS was used to check the amount of gold deposited on the samples after the SXRD and XSW measurements, and is given in equivalent monolayers of Au(111). Characterization of the samples by AFM was carried out before measurements by SXRD, and a typical AFM image is shown in figure 8. Au was deposited from 2M NaOH +0.1M $KAu(CN)_2$ + 0.2M NaCN solution in a rotating disk electrode configuration with a rotation speed of 1750 rotations min^{-1}. The deposition sequence was as follows: a first potential step to -2V_{mse} for 0.1s (low island density) or 2s (high island density) and a second step to -1.5V_{mse} for 20, 60 or 100s to produce different coverages of Au. At this deposition potential, the islands are seen to decorate the steps exclusively, and there is a significant distribution of particle sizes. The islands are tetrahedral (flat-topped triangular pyramids), and with AFM one can identify clearly 2 preferred orientations of the triangular base. The size distribution, and orientation can be studied in more detail using x-rays, and preliminary measurements have been carried out on 4 samples, with different Au coverage, and island density.

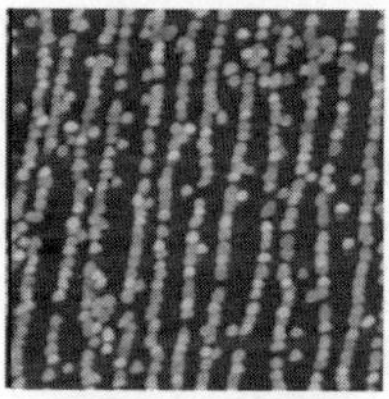

500nm x 500nm

Figure 8
Au deposited on 0.2° miscut Si(111)-H. AFM image of medium island density equivalent to 5 Au ML coverage.

Potential controlled Au deposition– SXRD

The SXRD measurements were performed in air, at beamline ID32, ESRF, using a photon energy of 10 KeV (λ=1.24Å). Each sample was aligned, and the co-ordinate system chosen such that the steps edges were oriented parallel to [-1,2,0]$_{hex}$ and the H axis, i.e., [1,0,0]$_{hex}$ is the direction descending the steps. Surface coordinates are again given in the hexagonal notation. Just as for the electroless deposition described previously, the Au clusters deposited under potential control are found to grow epitaxially with the Si(111) substrate, and have a preferred in-plane orientation along the high symmetry axes of the Si surface.

A significant difference in these samples as compared with the electroless deposited samples is the effect of the miscut Si(111) substrate on the symmetry of the surface. For each sample, a scan of the Au(111) peak was made both the q vector perpendicular to the steps (i.e. $\mathbf{q} \parallel [1,0,0]_{hex}$), and parallel to the steps (i.e. $\mathbf{q} \parallel [-1,2,0]_{hex}$, and is shown in figure 9a and b. Such measurements are in principle sensitive to three factors: the size distribution of the islands, the shape of the islands, and the correlation length between islands. A marked anisotropy is observed, which can be related to the well-correlated distribution of islands at the step edges, as opposed to *along* the length of the steps, where the correlation is less significant. Along the steps there may also be a contribution from the particle size distribution of the islands, a full fitting analysis is currently underway. Here, in addition to the AFM information, the x-ray analysis provides a statistical average of the size distribution and detailed, quantitative information about the shape of the smaller islands.

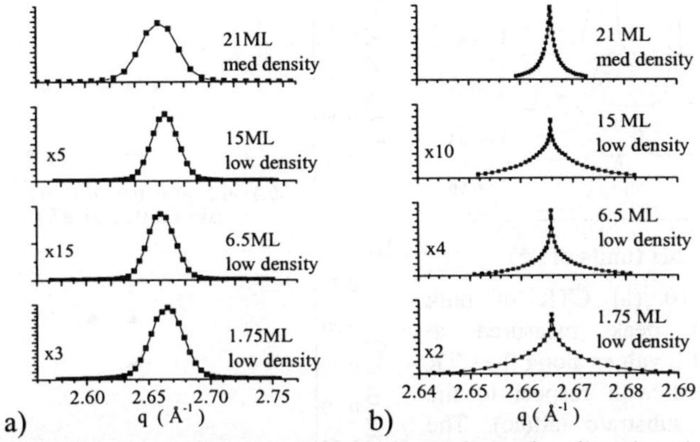

Figure 9 Scan of the Au(111) bulk peak in the direction (a) perpendicular to the steps edges and (b) along the steps edges for samples of different Au coverage and island density

A series of additional CTR peaks has been observed at an angle to the Au CTRs, which may indicate faceting of the islands (figure 10). The angles of these peaks are found to be different across the steps (around 30°), in comparison with the other 3-fold symmetry directions, such as along K, where the angle is around 40° from the surface normal. Data analysis is underway to fully understand these features and the variation in intensity shown at different L values. The influence of the miscut surface on the morphology of the gold islands, and how this varies with island density or size is also being studied.

Potential controlled Au deposition – XSW

XSW measurements were performed at around 12.5 KeV, (above the Au L edge at 11.9 KeV) in order to measure the standing wave modulation of the Au L fluorescence from the Si(111) reflection (perpendicular to the substrate surface). A sample with 0.21ML Au was measured, and showed no evidence for ordering of the Au on the Si miscut surface, in contrast to the samples produced by electroless deposition of gold.

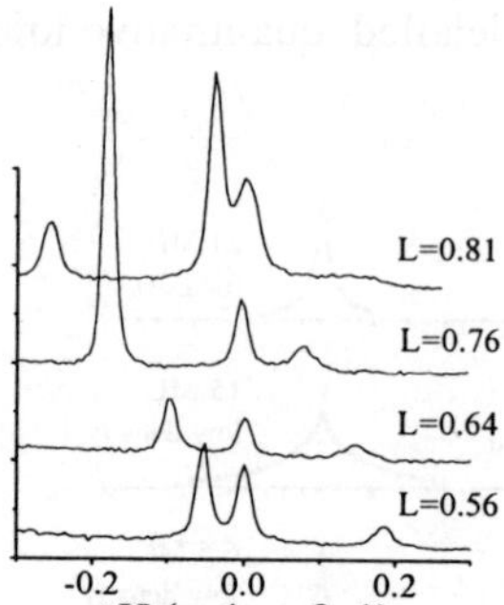

a)

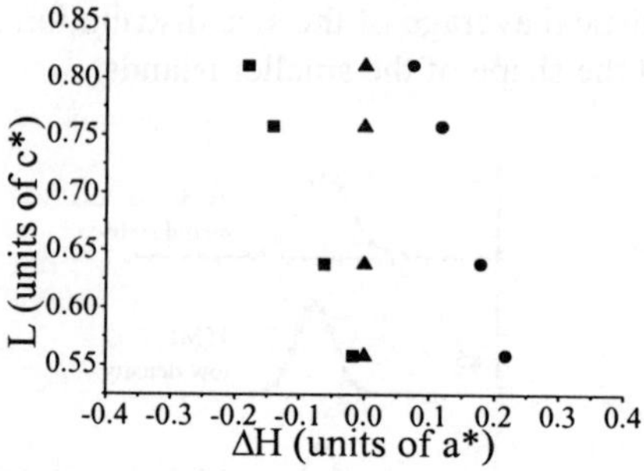

b)

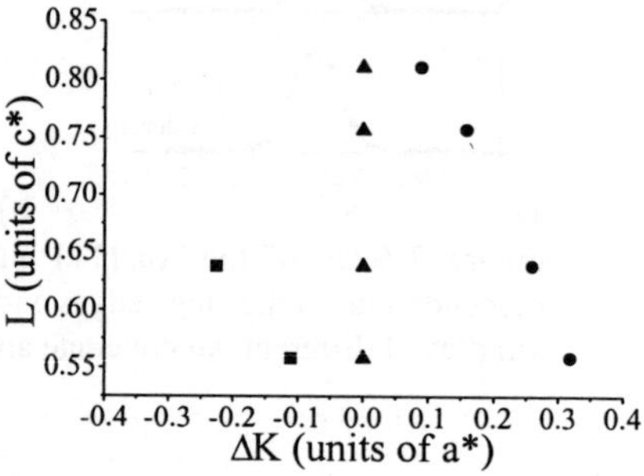

c)

Figure 10 (a) CTR of bulk Au(111) peak measured at different L values along the CTR (measured with respect to the Si(111) substrate lattice). The epitaxial Au(111) peak is at ΔH=0. (b) Angle of the 'facet' peaks along H (across the steps) of around 30° (c) angle of the 'facets' along K direction of 40°

Potential controlled Au deposition – Discussion

The epitaxy and orientation of the electrodeposited Au clusters are comparable with that observed for the electroless deposition described above. However in contrast, this method produces a particle size distribution of islands, which can be varied with the applied deposition potential. The miscut substrate surface also has significant effects on the gold cluster distribution and morphology. XSW measurements indicate that no ordered layer is formed, even at low coverages, unlike the electroless deposition which forms a 'wetting layer' as well as 3D clusters on the hydrogen terminated Si(111) surface.

4. Summary and Conclusions

The *in-situ* and *ex-situ* x-ray analysis of semiconductor electrodes and the electrodeposition of metals have revealed several novel findings. For Cu on GaAs(001) and electroless deposited Au on Si(111):H, wetting layers have been identified by XSW. However, in contrast to electrodeposited (sub)monolayers on metal surfaces, no long range ordered phase has been identified. We suggest this is due to the much smaller surface diffusion coefficient on semiconductors. Epitaxially well ordered 3D islands were identified by SXRD for all of the electrodeposited metals, which is different from the behavior of room temperature deposited metals in UHV.

Acknowledgments

We are grateful to Dieter Plachke, MPI Stuttgart, for recording and analysing the RBS data. We would like to thank the staff at HASYLAB, ESRF and NSLS for assistance with the synchrotron radiation measurements. Part of this work was supported by the German BMBF (under contract SE8GUA5). AR gratefully acknowledges the receipt of a stipend from the Fonds der Chemischen Industrie.

References

1. D.M. Kolb in *Advances in Electrochemistry and Electrochemical Engineering*, ed. H. Gerischer (Wiley, New York 1978) vol. 11, p125.
2. *Synchrotron Techniques in Interfacial Chemistry*, eds. C.A. Melendres and A. Tadjeddine, NATO ASI Series, C432 (Kluwer Academic, Dordrecht, 1994).
3. *Electrochemical Interfaces: Modern Techniques for In Situ Interface Characterization*, ed. H. Abruña (VCH, Weinheim, 1991).
4. R. Feidenhans'l, Surf. Sci. Rep., **10**, 105 (1989).
5. J. Zegenhagen, *Surf. Sci. Rep.*, **18**, 199 (1993).
6. W. Drube, H. Schulte-Schrepping, H.-G. Schmidt, R. Treusch, G. Materlik, *Rev. Sci. Instrum.* **66,** 1668 (1995).
7. J. Zegenhagen, A. Kazimirov, G. Scherb, D.M. Kolb, D.-M. Smilgies, R. Feidenhans'l, *Surf. Sci.*, **352-354** 346 (1996).
8. G. Scherb, A. Kazimirov, J. Zegenhagen, *Rev. Sci. Instrum.* **69**, 512 (1998).
9. R.W. Bernstein, A. Borg, H. Husby, B.O. Fimland, J.K. Grepstad, *Appl. Surf. Sci.*, **56-58**, 74 (1992).
10. U. Resch, N. Esser, Y.S. Raptis, W. Richter, J. Wasserfall, A. Förster, D.I. Westwood, *Surf. Sci.*, **260/270**, 797, (1992).
11. U. Resch-Esser, N. Esser, D.T. Wang, M. Kuball, J. Zegenhagen, B.O. Fimland, W. Richter, *Surf. Sci.*, **352-354**, 71 (1996).
12. G. Scherb, A. Kazimirov, J. Zegenhagen, T. Schultz, R. Feidenhans'l, B.O. Fimland, *Appl. Phys. Lett.*, **71**, 2990 (1997).
13. G. Scherb, D.M. Kolb, J. Electroanalyt. Chem. **396**, 151 (1995).
14. J. Zegenhagen, A. Kazimirov, G. Scherb, D.M. Kolb, D.-M. Smilgies, R. Feidenhans'l, *Surf. Sci.*, **352-354**, 346 (1996).
15. D.-M. Smilgies, R. Feidenhans'l, G. Scherb, D.M. Kolb, A. Kazimirov, J. Zegenhagen, *Surf. Sci.*, **367**, 40 (1996).

16. G. Scherb, A. Kazimirov, J. Zegenhagen, T.L. Lee, M.J. Bedzyk, H. Noguchi, K. Uosaki, *Phys. Rev. B*, **58**, 10800 (1998).
17. H.E. Hessel, A. Feltz, M. Reiter, U. Memmert, R.J. Behm, *Chem. Phys. Lett.* **186**, 275 (1991).
18. I.K. Robinson, *Phys. Rev. B*, **33**, 3830 (1986).
19. J.C. Ziegler, G. Scherb, O. Bunk, A. Kazimirov, L.X. Cao, D.M. Kolb, R.L. Johnson, J. Zegenhagen, *Surf. Sci.*, **452**, 150 (2000).
20. G. Oskam, J.G. Long, A. Natarajan, P.C. Searson, *J. Phys. D*, **31**, 1927 (1998).
21. G. Oskam, J.G. Long, M. Nikolova, P.C. Searson, *Mat. Res. Soc. Symp. Proc.* **451** (1997), 257.
22. S. Warren, A. Reitzle, A. Kazimirov, J.C. Ziegler, O. Bunk, L.X. Cao, F.U. Renner, D.M. Kolb, M.J. Bedzyk, J. Zegenhagen, *Surf. Sci.* **496**, 3, 287-298 (2002).
23. C. Rossiter, I.I. Suni, *Surf. Sci.*, **430**, L553 (1999).
24. R. Srinivasan, I.I. Suni, *Surf. Sci.*, **408**, L698 (1998).
25. R. Srinivasan, I.I. Suni, *J. Electrochem. Soc.* **146** (1999), 570.
26. P. Allongue, C. Henry de Villeneuve, S. Morin, R. Boukherroub, D.D.M. Wayner, *Electrochimica Acta*, **45**, 4591-4598 (2000).
27. M. Munford, R. Cortès, P. Allongue, *Sensors and Materials*, **13**, 259-269 (2001).

ASPECTS OF HETEROEPITAXIAL GROWTH

S.M. SHIVAPRASAD

Surface Physics Group

National Physical Laboratory, New Delhi– 110 012, India

E-mail: prasad@csnpl.ren.nic.in

The paper discusses several aspects of heteroepitaxial growth including bonding, lattice mismatch and surface free energy, via several in-situ UHV experiments. The Ni/Ru(0001) shows a novel pathway for the misfit strain relaxation by the formation of a quartet reconstruction, while the atomically rough W(111) shows a massive faceting with an adsorbed metal monolayer. In case of the Sb/Si(111) interface, the dangling bond minimization schemes result in a delayed onset of band bending and a variety of novel surface phases. The results demonstrate the dominant role the kinetics of growth plays in determining the morphological, structural and electronic properties of interfaces.

1. Introduction

The Low dimensional structures demonstrate properties that depend on their size and shape and completely different from their bulk behavior. When the dimensions are small enough for quantum effects to prevail, exotic magnetic, catalytic and electronic properties are manifested. However, forming controlled, ordered and reliable nanostructures of metals and semiconductors is still challenging. One of the near ideal routes, is the attempt to form ultra thin films in ultra high vacuum conditions, which provide the best handle on the control and study of the initial stages of growth. The surface processes involved in the initial stages of growth have been extensively dealt with in literature [1,2]. Thermodynamic arguments have been used to explain the role of the surface free energies along with the effective interfacial energies in determining the growth modes, which are generally of three types: a) the layer-by-layer or Frank van der Merwe b) island or Volmer-Weber and c) the layer plus island or the Stranski-Krastanov mode. However, in 'real cases' it is observed that issues are more complicated than the mere thermodynamic and statistical mechanical models.

Film growth can be viewed as the sequential arrival of atoms from the vapor (or a liquid) onto a substrate, which at the outset can re-evaporate or stick onto the substrate, creating an adatom density. These adsorbed adatoms now attempt to occupy the potential energy minima sties, while they make vibratory attempts to escape the potential wells. Depending on the energetics, the adatom may diffuse on the surface being associated with a thermally activated diffusion energy E_d. These migratory adatoms in search of lower potential energy sites may pre-maturely encounter other diffusing adatoms (to form dimers or larger clusters) or special sites such as defects, steps, etc or even undergo intermixing with the substrate. The clusters after attaining a critical size are immobilized and consequently grow in size to eventually coalesce into larger 2-D or 3-D islands. The evolution of the shape and size of the islands is a consequence of self-organization driven by minimization of surface free energy. Though most of these arguments are based on thermodynamics, the growth of thin epitaxial film is a non-equilibrium kinetic process in which one or more steps are rate limiting. For example a) the vapour pressure is dominated by the sublimation energy that is much larger than kT, below the melting point, b) the lattice vibrations influence strongly (as v^3) and, c) the state of the substrate, inter-diffusion, interactions, etc., cannot be ignored. In the thermodynamic picture the morphology of the growing film is determined by the balance between the free energies of the film surface, substrate surface and the interface between the two. In the kinetic picture the morphology is a result of the interplay of various atomic diffusion processes, active on a time scale set by the deposition flux and substrate temperature. The final structure of the film depends on many possible 'kinetic' effects each with its characteristic length and time scales. This constant competition between the thermodynamics and kinetics, complicate the prediction of film growth and consequently their properties, however, it also provides an unique opportunity to freeze systems in several novel surface phases with exotic characteristics and providing us the 'tailorability' of properties.

Among several aspects of ultra thin film growth in this presentation a few chemical aspects are discussed. In heteroepitaxial growth since two different materials are involved, it is obvious that strain, induced due to differences in surface free energies, lattice mismatches and nature of the interaction between the arriving adatoms and the substrate electronic structure, are crucial. Along with this several kinetic handles like deposition rate, substrate and annealing temperatures and presence of surfactants and defects play dominant roles. Examples from the author's experience that highlight each of the effects are presented here. However, it must be realized that though each system tries to stress on one dominant aspect, several others are also simultaneously operative.

2. Experimental

These experiments are performed *in-situ* in ultra high vacuum with several surface sensitive techniques complementing each other, in several laboratories. In vacuum systems with a base pressure of $\sim 10^{-11}$ torr, metals are adsorbed on specifically atomically clean substrates with sub-monolayer coverage control. In the Molecular Beam Epitaxy kind of growth, thin films are grown in experimental conditions far from equilibrium. The growth is monitored *in-situ* by Auger Electron Spectroscopy (AES) and X-ray Photoelectron Spectroscopy (XPS) and Electron Energy Loss Spectroscopy (EELS) for stoichiometry, growth modes, cleanliness of deposition and electronic property determination. Low Energy Electron Diffraction (LEED) and a particularly high resolution Scanning Tunneling Microscopy (STM) are employed to study the surface structures and morphology.

3. Results and Discussions

Three systems of epitaxial growth of metals on single crystal metal and semiconductor surfaces are discussed below:

3.1. *Ni/Ru(0001)*

In most cases of heteroepitaxial growth lattice mismatch between deposit and the substrate will produce a strain in the epitaxial layer, affecting the layer morphology and forcing 'metastable artificial' structures. The growing film initially conforms to the substrate geometry adopting a pseudomorphic structure. As the thickness increases so does the strain energy and the epitaxial film eventually can choose several pathways to relieve strain, typically by introduction of misfit dislocations or by surface roughening [3,4]. While the importance of strain relief on film structure and morphology has long been studied, the effect of strain on the kinetic processes has hardly been recognised. In case of the Ni/Ru system itself, layer dependent mobility of adsorbates, suggesting a dominant dislocation density in the first monolayer (which later reduces for higher thicknesses) has been reported [5].

The experiments are performed in an UHV chamber equipped with LEED, AES, STM and in-situ Ni evaporation from an electron bombardment source. The STM images were acquired from a particularly high resolution STM at tunnel currents of 1 nA. CO was pre-adsorbed on the substrate by backfilling the chamber using a fine-control needle valve.

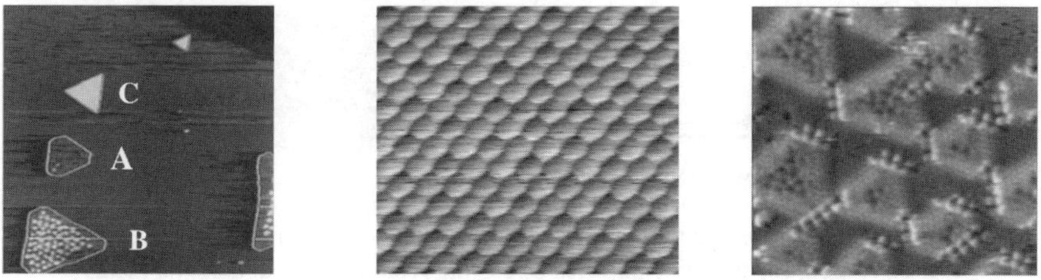

a b c

Figure 1. STM images of Ni islands on Ru(0001) surface. a) is the wide scan showing three types of islands formed. b) atomic resolution images of pseudomorphic Type A islands c) details of dislocated Type B islands.

Figure 1 shows the STM image for a 0.12 ML Ni adsorbed on a Ru(0001) substrate held at 380K at flux rate of 0.045 ML/min. The image clearly shows three kinds of islands marked A, B and C in the figure. Figure 1a shows high resolution images of A and B types of islands. Type A are pseudomorphic (1x1) Ni islands with truncated triangular shapes. Type B are larger than type A islands and are covered with bright features all over (except at the edges). These features by the lattice mismatch, which force atoms to change from hcp to fcc sites. Type C are the novel sharp triangular features that are raised by 0.2 Å. High resolution images of the type C islands show that they have a (2x2) reconstruction. It is also interesting to note that the direction of these islands is opposite to the types A and B. Since these islands appear only in the presence of pre-adsorbed CO, the percentage density of the (2x2) islands is plotted as a function of CO coverage in Figure 2a.

Now we probe for details into the Type C islands. High resolution images in Figure 2b show the (2x2) configuration as well as domain boundaries. However atomic resolution reveals that alternate Ni adatoms in alternate rows are raised by 0.2 Å, thus resulting in the (2x2) structure. Also making a statistical measurement between the centers of the atoms observed, it is clear that there are groups of 4 Ni atoms each, which we call as a 'quartet reconstruction'[6]. The raised atom occupies the three-fold hollow fcc site, while in the attempt to form the quartet, the other Ni adatoms bunch together and thus raised from their fcc site positions, since the Ni bulk inter-atomic distance (2.49Å) are smaller than those of the substrate Ru (2.8Å). The fourth atom is constrained to move the maximum in the z-direction and thus is seen as the brightest in the STM images.

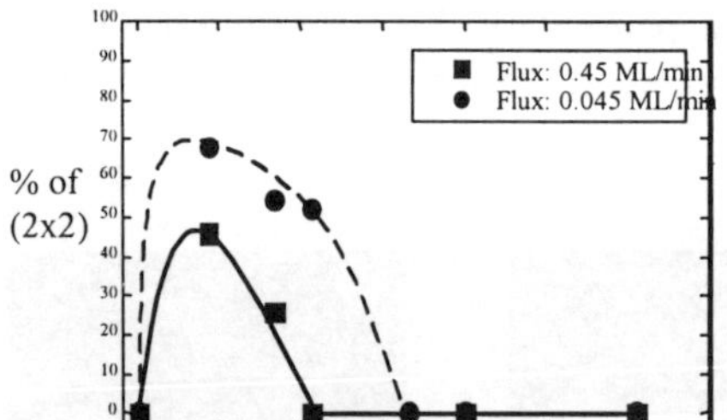

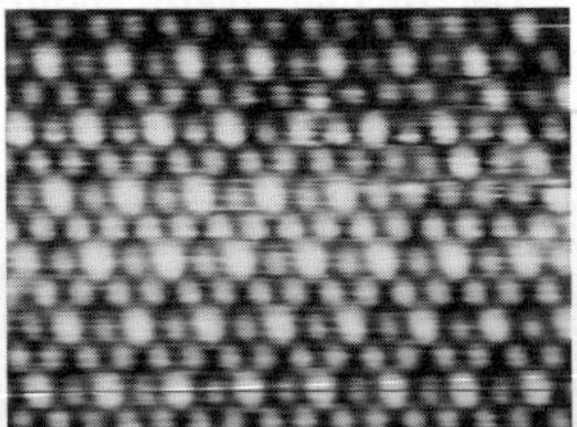

Figure 2. a) Shows the percentage of (2x2) islands as a function of pre-adsorbed CO coverage, for two Ni flux rates b) atomic resolution image of the (2x2) 'quartet reconstruction'.

The fact that the (2x2) island is rotated with respect to the type A and B islands suggests that either the atoms have occupied the fcc sites instead of the bulk stacking hcp positions, or are made of B-steps. Looking at regions where we observe the lateral interface between the (1x1) and (2x2) phases, we can clearly see that there is a difference in the site occupation between the two configuration islands. Now observing the direction of islands that are formed at the step edges (bulk stacking), we see that the (1x1) islands (Type A and B) have Ni atoms at hcp sites of the substrate. Thus it is clear that the Ni atoms in the type B (2x2) islands occupy the fcc substrate sites.

It is important to know the possible mechanism in which CO plays a crucial role to effect this reconstruction. It is known that CO adsorbed on Ru substrates occupy the hcp sites. This restricts the adsorption of the initial Ni adatom to the fcc sites. Now similarly when the other nearby fcc sites, which have a lower potential energy barrier in comparison to the hcp sites, an attractive interaction between the Ni adsorbates becomes possible. However the subtle resultant causes the 'quartet reconstruction'. Figure 2a shows the fact that only small amounts of CO is necessary for the formation of the (2x2) islands, but the CO molecules do not integrate into the island. Thus the (2x2) quartet reconstruction is a novel pathway for the strain relaxation with energy requirement intermediate to the pseudomorphic and the misfit dislocated situations. The CO acts as a mobile surfactant that induces the occupation of the fcc sites for the Ni adatoms [7]. This configuration rotates the direction of the islands and maintains the triangular shape of the islands. These (2x2) islands are seen to dislocate at larger coverages due to incorporation of more adatoms. This mode of strain relaxation can have general implications for understanding the influence of kinetics on hetero-epitaxial systems.

3.2. *Sb/Si(111)*

Semiconductor surfaces and interfaces display remarkable electronic and structural properties specifically due to the existence of surface and interface states [8]. Bulk termination of solids at surfaces result in dangling bonds that create surface states and on adsorbate-covered surfaces from bonds between adsorbate and semiconductor surface atoms. For metal adsorbates, the wave-function of the metal electron decay exponentially into the semiconductor,

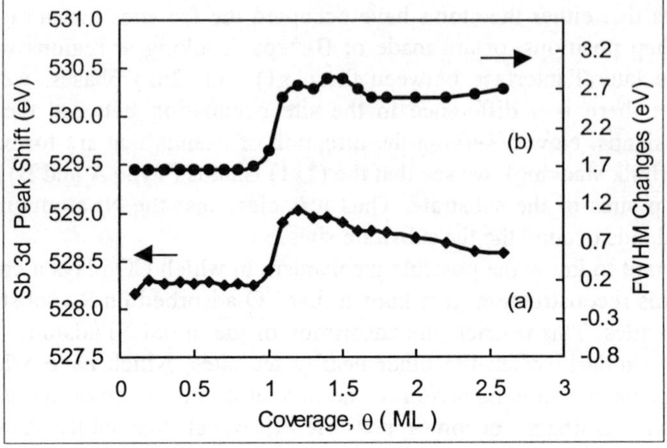

Figure 3. Plot showing the position shift in the Sb 3d core level along the y-axis, and the FWHM changes along the alternate y-axis, as a function of Sb coverage.

charging the surface and interface states. Adatoms and surface reconstructions reduce the density of the dangling bonds and thus the total energy of the surface. Adsorbates form chemical bonds with semiconductor adatoms and saturate the dangling bonds. In the ideally terminated Si(111) surface of Si, each surface atom would have one dangling bond which contain one electron. The respective dangling bond levels are below the valence band maximum so that, in thermal equilibrium they should be occupied by two electrons. An electron transfer from one half of the surface atoms to the other half might achieve this. Such charge re-distribution would require a reconstruction and a re-hybridization of bonds in the selvedge which result in the severe surface strain. The driving mechanism for the adsorbate induced reconstructions is the saturation of the dangling bonds and as a result of this, a lowering of surface energy, an effect which is counteracted by adsorbate induced strain. Experimental results seems to indicate that Group V atoms cause ($\sqrt{3} \times \sqrt{3}$)-R30° reconstruction rather than a (1×1) structures (as in As) when the covalent radii of the adsorbate becomes larger by a factor of 1.08 than substrate atoms. Intuitively this seems to be plausible since the bonds between adatoms and surface atoms of the substrate will be the more distorted, larger their difference in size is. A change in bond length by 8% corresponds to an increase of the tetrahedral bond angle by 9.7%.

Sb is evaporated from a home-made Ta K-cell at a low flux rate onto a clean (7x7) reconstructed Si(111) surface held at room temperature (RT). Plotting the uptake curve (not shown here) of the Sb(*3d*)/Si(*2p*) intensity ratio versus the deposition time, shows two linear segments characteristic of the Frank van der Merwe growth mode. The break in slope is due to the attenuation of the core level electrons from the first monolayer due to scattering from the second layer adatoms. From this, and correlating with the XPS ratio obtained for a 1ML Sb adsorbed on a 540°C substrate yielding a ($\sqrt{3} \times \sqrt{3}$)-R30° structure [9], we obtain a calibration of our deposition rate as 0.03 ML/min. With suitable techniques of background subtraction and smoothing, the core level peaks are deconvoluted into Gaussian components to identify the peak positions and widths. Figure 3 shows the coverage dependence of the Sb($3d_{5/2}$) peak shifts and the FWHM along the alternate y-axes. Figure 4a shows the results from the LEED studies including the patterns observed during the uptake and the intensity ratio of the (1x1) integral spots and the $1/7^{th}$ fractional order spots, as a function of Sb coverage. It is clear from Figure 3 that at a coverage of 1 ML there is a sharp shift as well as broadening of the core level Sb peak [10]. The LEED studies show a sharp decrease in the clean Si (7x7) characteristics (Fig. 4b) around 1 ML coverage, resulting in a (1x1) surface phase (Fig. 4c). ELS studies measuring the volume and surface plasmon peak positions and surface states related transitions clearly indicate (now shown here) that the surface related features also show significant changes at 1 ML, while the bulk related peaks show monotonic behavior over the coverage range studied.

Discarding the shift in core-level Sb3d peak to be due to a chemical shift, which is highly unlikely at these temperatures for this immiscible system, we attribute it to be entirely due to band-bending. To understand the delayed onset of band-bending seen in the XPS data, we resort to the LEED data which suggests that the (7x7) reconstruction of Si persists up to a coverage of 1.0 ML, around which it transforms into a (1x1) phase. Consulting the DAS (Dimer-Adatom-Stacking Fault) model (8) that describes the atomic arrangement of the (7x7) reconstructions, it is realised that the Si adatoms give rise to a surface state

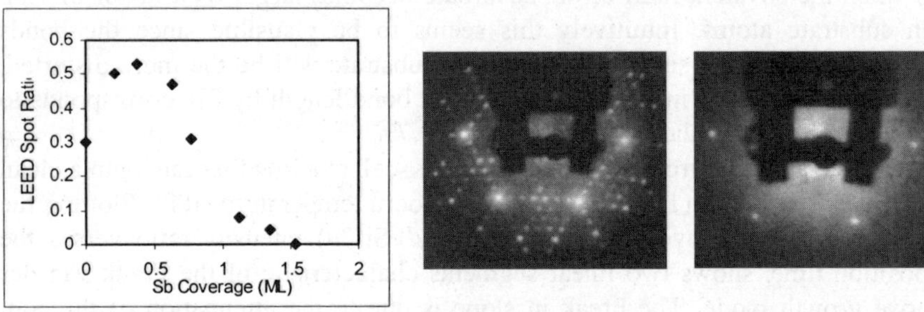

Figure 4. a) shows the LEED intensity ratio of the integral and fractional order spots. b) LEED pattern of clean Si (111) (7x7) surface c) pattern for the (1x1) surface obtained after 1 ML Sb coverage.

at about 0.25 eV, which is observed to disperse across the Fermi level and thus rendering the reconstruction metallic. Thus, as long as the (7x7) phase of the substrate remains intact, band-bending is not observed. However, when the metallic surface phase transforms into a (1x1) semiconducting phase, we see a sharp onset of the barrier, and also a significant broadening of the core-levels. Thus the slow flux rate ensure the initial Sb adatoms to occupy conformal (1x1) sites sequentially on both the unfaulted and faulted halves of the (7x7) unit cell. As the coverage increases, due to availability of sufficient adatom charges, the strain in the reconstructed layer is gradually relaxed, to result in a (1x1) phase. Initially the dimers and stacking faults would be rearranged but finally when the Sb coverage approaches 1 ML, the Si adatoms are freed to migrate on top of the Sb layer and fill the corner holes of the DAS reconstruction and consequently lifting it.

Figure 5a shows the phase diagram that indicates the temperature dependence of the Sb overlayer atomic arrangement on the Si substrate. Under several minimization schemes, several surface phases are observed. It is to be

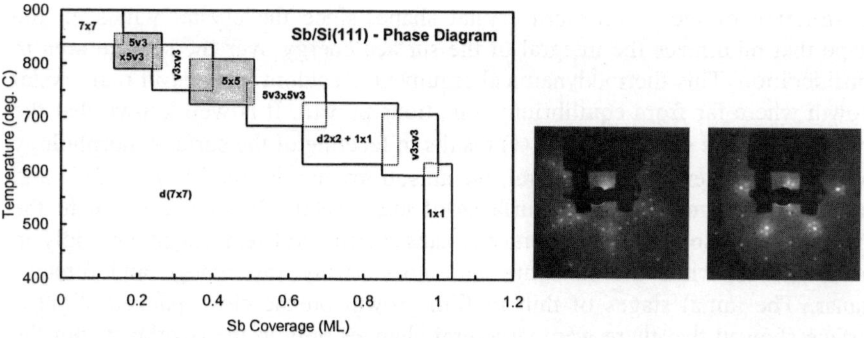

Figure 5. a) shows the phase diagram of the Sb/Si(111) system. b) and c) show the novel (5x5) at 0.4 ML and a (5√3 × 5√3) at 0.2 ML phases, respectively.

noticed that the (√3 × √3) at 1.0 and 0.33 ML, d(2x2) at 0.8 ML, (5√3 × 5√3) at 0.6 ML and the (7x7) at 0.1 ML are observed when Sb is adsorbed onto a clean Si(111) 7x7 surface held at the respective temperatures. However, if Sb is adsorbed at room temperature giving the (1x1) phase is annealed, then the pathway adopted during desorption is different and yields novel (5x5) at 0.4 ML and a (5√3 × 5√3) at 0.2 ML phases [11], whose LEED are shown in Fig.5b and 5c, respectively patterns.

The study clearly shows how adsorption at room temperature at low flux rates can enable different adsorption sites for the Sb adatoms on the (7x7) reconstructed Si(111) surface. Adsorption at higher temperatures which provide more energy and mobility to the adatoms can relax the reconstruction induced strain while thermal desorption of the room temperature formed system adopts an entirely different route of passivating the substrate dangling bonds [12]. However, it would be interesting to understand the actual arrangement of the atoms in these phases, which can lead to tracing the mechanism of minimization of dangling bonds in the system.

Other results regarding the observation of surface phases formed by varying the kinetics of growth in the Ag/Si (111) and (100) and Mn/Si(111) have been studied and several new phases observed, and are reported elsewhere [13-18].

3.3 Pt/W(111)

The surface tension of a planar solid depends on the crystallographic orientation of the sample[19]. A polar plot of the surface tension (surface free energy) with the crystallographic orientation is given by the famous Wulff's flower plot. The anisotropy of the surface free energy essentially determines the construction of the equilibrium crystal shape, since the crystal will seek the shape that minimizes the integral of the surface energy over the surface area in consideration. This thermodynamical argument is seldom realized in real crystal growth where far from equilibrium conditions prevail. It is well known that the anisotropy of the surface energy (γ) results in faceting of the surface morphology to minimize the energy. In general, the anisotropy in γ is small [$\Delta\gamma/\gamma < 3\%$] and stable for clean close packed surfaces of pure metals. In such cases with the adsorption of another metal overlayer, causes structural rearrangements only in the overlayer, while the substrate undergoes hardly any change and remains planar. The initial stages of thin Pt film growth on the close packed W(110) surface showed that there were structural changes only in the overlayer, but the substrate remained planar. Thermal desorption studies showed [20,21] that molecular CO is more weakly bound on a Pt/W(110) surface than either to bulk Pt or clean W substrate, demonstrating that the structural changes in the multilayer films result in a W surface that is mostly covered by a monolayer Pt film with unique CO chemisorption properties.

a) b)

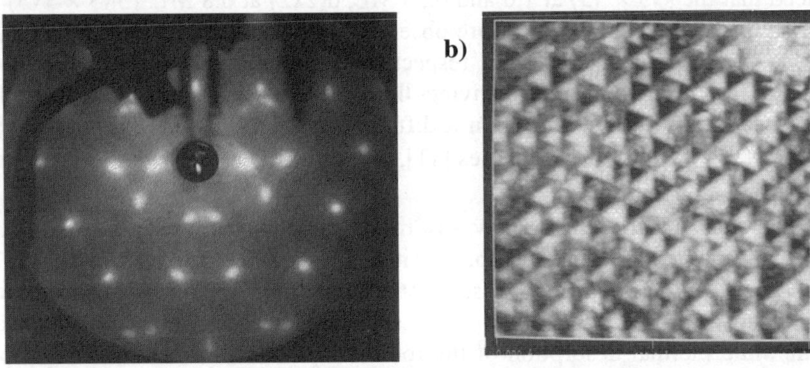

Figure 6. shows the LEED (a) and the STM image (b) of the Pt induced pyramidal faceted surface of W(111).

However, the scene is different when we consider crystallographic faces that are atomically open and rough. Here the example of the (111) surface of the body centered cubic metal W, with an atomically rough surface with high surface energy is considered. Single crystal W(111) samples are heated resistively or by electron bombardment and the sample cleanliness is verified in-situ by AES. Pt wire wound on a W filament is the source of Pt evaporated onto the substrate. Coverages are estimated by a combination of AES and Thermal Desorption Spectroscopy of the metal overlayers. In simple kinematic LEED theory, as the beam energy is increased, the diffracted beams should converge to the specular

(0,0) beam. Pt of coverage > 1ML was dosed onto the clean W(111) surface and annealed to 1100 K. This results in a LEED study where the non-specular beams are not found to converge onto the (0,0) beam, but move in three directions away from the macroscopic normal, as the beam energy is increased. The LEED pattern is shown in Fig.6a. This indicates the formation of new specular directions, since, for a planar surface the diffraction beams move inward converging on to the specular (0,0) beam, as the electron beam energy is increased, as a consequence of the structural transformation. These beams are identified to be due to diffraction from {211} oriented facets with (1x1) unit cells [22,23]. The size of the spots enables us to estimate an average facet dimension to be about 10 nm. This faceting has been studied by a scanning tunneling microscope operated in air. The STM clearly shows the formation of a dense network of triangular features on the surface (Fig. 6b). The STM images not only confirm the pyramidal nature of the facets (Fig. 7b), but also give information regarding the facet spatial distribution and sizes. As the system is annealed to various temperatures, the average facet size proportionately increased, reaching a value of 30 nm at an annealing temperature of 1200K.

CO is an excellent probe molecule to study thermal desorption spectroscopy. CO typically adsorbs molecularly at temperatures <350K, followed by a dissociative adsorption at higher temperatures, on clean W surfaces. For a Pt covered W surface, molecular adsorption increases, while the dissociatively adsorbed component decreases, since CO does not dissociate on bulk Pt [24]. To examine the relation between the observed structural changes and the CO reactivity on these surfaces, TDS experiments were performed on different surface structures. Fig.8 shows the TDS spectra in the region of molecular adsorption. Before annealing (planar morphology), it has two peaks corresponding to molecular adsorption on W(111) surface. However after annealing, we see that one of the peaks has disappeared, probably due to the agglomeration of the Pt multilayer adatoms. These results show that faceting influences the CO chemisorption properties significantly. The results provide evidence that a critical thickness is necessary to cause facets, but this may be happening at a much lower temperature than that observed by the LEED

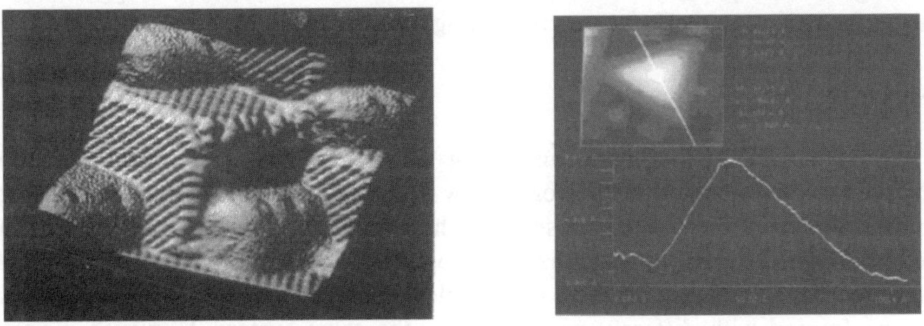

Figure 7. Atomic resolution image of the faceted surface showing the pyramidal arrangement and its height profile.

experiments which need a large enough coherence length. This was proved recently, by the observation of atomic resolution STM images of these facets, shown in Fig.7a.

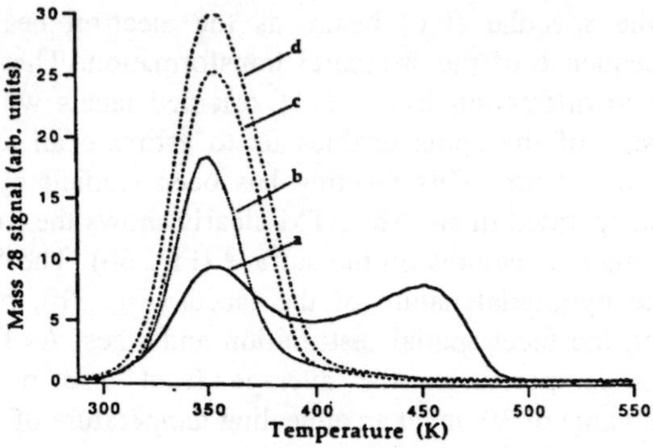

Figure 8. TDS spectra of molecularly adsorbed CO on a W(111) surface covered with 3.0 ML of Pt annealed to a)300K, b)550K, c)1200K and d)1600K

As discussed earlier, thermodynamical considerations suggest that the monolayer decreases the surface free energy of the system, but also increases the anisotropy of the surface energy as a function of crystallographic orientation. This factor can be is responsible for the monolayer metal induced faceting phenomenon. In a long series of experiments several ultrathin metal (Ti, Co, Ni, Cu, Ag, etc) monolayers on W(111) were studied, where neither the LEED or STM showed evidence for any faceting. However, for metals like Pt, Pd, Au, Rh, and Ir, the W(111) substrate reconstructs to form facets. O, S, Cl are also found to facet the open surface. Trying to correlate a particular bulk property that may be responsible, atom size, surface enery, elastic modulus, cohesive energy, Pauling electronegativity were considered. The only property that seemed to correlate with faceting of W(111) is seen to be for overlayer elements having a Pauling electronegativity values greater than 2.0. This correlation suggests that the surface electronic effects are controlling the structural instability of W(111).

4. Conclusions

Several UHV experiments including adsorption and in-situ probing with surface sensitive techniques performed by the author and collaborators relating to heteroepitaxial growth are presented. The studies show the role of kinetics in determining the overlayer structure and properties, by adopting various pathways of surface strain relaxation. The results clearly show that several novel phases can be formed with different properties for the same system. However, a very systematic and copious amount of study is necessary to understand the actual control that can help tailoring desired structural, chemical and electronic properties of nanostructures.

Acknowledgements

The author wishes to acknowledge several research colleagues from National Physical Laboratory, India, Rutgers University, USA and University of Ulm, Germany where respective parts of the experiments were conducted. Special thanks are to Prof. T.E. Madey, Prof. R. J. Behm and the Director, National Physical Laboratory, for their guidance and encouragement. Thanks are also due to Mr. Vinod Paliwal for hekp in preparing the manuscript. Research grants from Deptt. of Science & Technology, India are gratefully acknowledged.

References

1. M. Hanbucken, M. Futamoto and J.A. Venables, *Surf. Sci.* **147,** 433 (1984)
2. E. Bauer, Z. Kristallogr. **110,** 372 (1958)
3. F. Frank and J.v.d. Merwe, *Proc. Roy. Soc. Lond.* **A198**, 205 (1949).
4. R.J. Asaro and W.A. Tiller, Metall. Trans. **3,** 1789 (1972)
5. J.A. Meyer, P. Schimd and R.J. Behm, *Phys. Rev. Lett.* **74,** 3864 (1995).
6. S.M. Shivaprasad, Y. Shingaya and R.J. Behm, Submitted for publication.
7. H.A. van der Vegt, *Phys. Rev. Lett.* **68,** 3335 (1992)
8. W. Monch, "Semiconductor Surfaces and Interfaces",Springer-Verlag, Berlin, (1995).
9. H.B. Elswijk, D. Dijkkamp, E.J. Van Loenen, *Phys. Rev.* **B44,** 3802 (1991)
10. V.K. Paliwal, A.G. Vedeshwar and S.M. Shivaprasad, *Pure & Appl. Chem.* **74**, (2002) In press.
11. V.K. Paliwal, A.G. Vedeshwar and S.M. Shivaprasad, *Surf. Sci. Lett.* **513**, L397 (2002)
12. V.K. Paliwal, A. G. Vedeshwar and S.M. Shivaprasad, *Phys. Rev.* **B 66**, (2002) 245404
13. S.M. Shivaprasad, Sandeep Singh and Y. Aparna, *Solid State Commun.* **107**, 257 (1998)
14. S.M. Shivaprasad, T. Abukawa, H.W. Yeom, M. Nakamura, S. Suzuki, S. Sato S. Sakamoto and S. Kono, *Surf. Sci.* **344**, L1245 (1995)
15. M. Shimomura, T. Abukawa, M. Higa, M. Nakamura, S.M. Shivaprasad, H. W. Yeom, S. Suzuki, S. Sato, J. Tani and S. Kono, *Rev. Surf. Sci.* **5**, 953 (1998)
16. S.M. Shivaprasad, C. Anandan, S.G. Azatyan, Y.L. Gavriljuk and V.G. Lifshits, *Surf. Sci.* **382**, 258 (1997)
17. Sandeep Singh, S.M. Shivaprasad and C. Anandan, *Appl. Surf. Sci.* **152**, 213 (1999)

18. Sandeep Singh and S.M. Shivaprasad and C. Anandan, *Appl. Surf. Sci*. **136**, 189 (1998) A. Zangwill, "Physics at Surfaces", Cambridge University Press, New York, (1990)
19. S.M. Shivaprasad, R.A. Demmin and T.E. Madey, *Thin Solid Films*, **163**, 393 (1988)
20. R.A. Demmin, S.M. Shivaprasad and T.E. Madey, *Langmuir*, **4,** 1104 (1988)
21. K.J. Song, .Z. Dong and T.E. Madey, *Langmuir,* **7**, 3019 (1991)
22. T.E.Madey, Jie Guan, C.Z. Dong and S.M. Shivaprasad, *Surf. Sci.* **287/288**, 826 (1993)
23. C.Z. Dong, S.M. Shivaprasad, K.J. Song and T.E. Madey, *J. Chem. Phys*. **99**, 9172 (1993)

QUANTUM WELL, WIRE AND DOT: STRUCTURE AND TRANSPORT

GROWTH AND CHARACTERIZATION OF P-HEMT STRUCTURES GROWN BY MOLECULAR BEAM EPITAXY

R.MURALIDHARAN , T SRINIVASAN ,UMESH TIWARI ,S.K.MEHTA AND R.K. JAIN

Solid State Physics Laboratory Lucknow Road, Delhi-110054,India

D. V. SRIDHARA RAO, K. MURALEEDHARAN, R.BALAMURALIKRISHNAN

Defence Metallurgical Research Laboratory, Hyderabad

In this paper issues involved in the growth of high quality pseudomorphic High Electron Mobility Transistor (P-HEMT) structures are discussed. Procedures adopted for the growth and characterisation of various constituent layers of P-HEMT are described. Results on the layer and device characteristics are presented

1 Introduction

Pseudomorphic High Electron Mobility Transistors (PHEMTs) based on AlGaAs/InGaAs/GaAs structures are being widely used as active devices for fabrication of millimeter wave circuits [1]. It is well known that high carrier mobility, high sheet electron concentration, low gate to source resistance, low gate-source capacitance, good isolation between devices and low capacitance of the buffer layer are important parameters for obtaining good characteristics of HEMTs. In this paper, we describe some of the material related issues which influence the device performance and also present results on the growth and characterization of P-HEMT structures grown on low temperature (LT) grown GaAs buffer.

The basic issues involved in the growth of PHEMT structure are (a) buffer layer quality (b) pseudomorphic active channel layer with uniform Indium composition and low dislocation density (c) AlGaAs barrier layer with

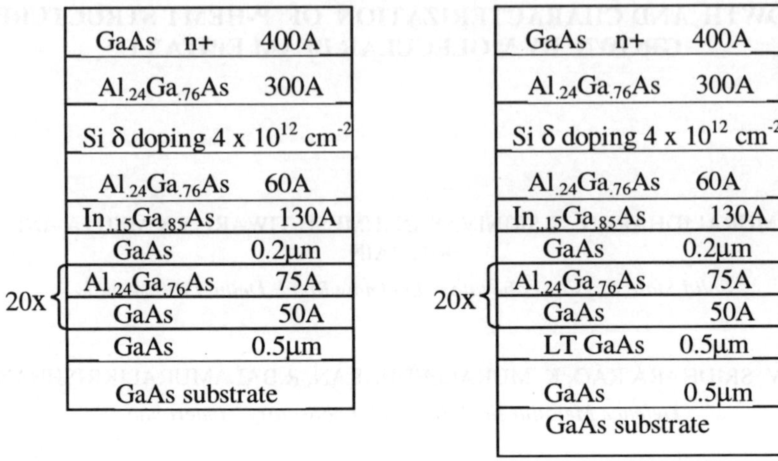

Figure 1. Typical PHEMT structure (a)Normal GaAs with SL buffer (b) LT GaAs with SL buffer

low defect concentration and (d) Si δ doped layer with minimum spread of dopant concentration. We shall discuss the growth and characterisation related aspects of each of these layers in the following sections

2 Experimental

We use a RIBER 2300 R & D molecular beam epitaxy (MBE) system with solid sources for the growth of epitaxial layers. Substrate temperature was monitored using thermocouple and was calibrated using the substrate oxide desorption temperature which is taken as 580 C. We have used both indium free as well as indium mounted molybdenum holders for mounting the substrate. As_4/Ga Beam Equivalent Pressure (BEP)was kept around 20 and the growth was carried out under arsenic stabilised conditions.
Photoluminescence was carried out at 17K using the Ar ion laser as the excitation source and 0.25m monochromator for dispersion and GaAs photo multiplier tube /Si diode detector for detecting the emitted radiation. Hall measurements were carried out in Vander Pauw geometry with "In" alloyed ohmic contacts . High Resolution X ray Diffraction (HRXRD) was performed in Philips Xpert system. The cross-sectional transmission electron microscope (TEM) specimen was prepared by argon ion milling using Gatan's Duomill. The conditions employed for ion milling were: double sided ion beam milling with 12-degree gun angles, individual gun current of 0.5mA, liquid nitrogen cold stage for specimen cooling and IOLAR-2 grade argon. The (200) dark field

Cross section TEM image, shown was observed at 250kV near [011]zone axis using Philips EM430T analytical transmission electron microscope

2.1 *Buffer layer Growth*

Generally a buffer layer is grown on the substrate before any active device structure is grown. The purpose of this layer is to prevent/trap defects and impurities, which are present on the semi-insulating substrate surface, reaching the active layer. However, the electrical and structural quality of the buffer layer is extremely important . In practice, it is found that if the background doping is 10^{15} cm $^{-3}$ or higher, device isolation is not very good even though the device performance is not necessarily affected. Further, the buffer layer quality contributes to back gating effects, which are also critically dependent on the resistivity of the buffer layer. In order to overcome this problem, growth of LT GaAs buffer layer has been used [2]. As grown LT GaAs contains a large amount of excess "As" which could be up to 1-2%. The presence of excess "As" in the GaAs lattice produces tetragonal distortion. We have estimated the amount of excess "As" in LT GaAs grown at 300 C.in the following manner.

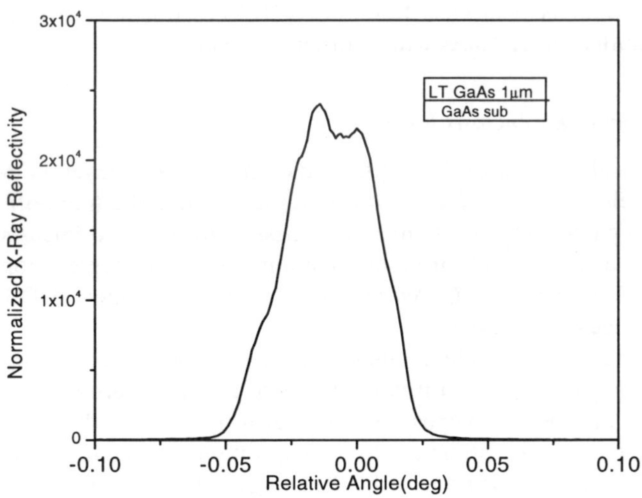

Figure 2. Rocking curve of the 400 reflection of LT GaAs grown at 300C

X ray rocking curve of data (shown in Fig 2) of as grown LT GaAs grown at substrate temperature of 300C was recorded with a 4-crystal Ge (220) incident beam monochromator .This contains many peaks due to non-uniform distribution of excess "As" in the layer[3] .However we have taken the extreme peak separation viz .022deg and calculated

$$\frac{\Delta d}{d} = -\Delta\theta Cot\theta$$

and using the value of ε =0.3 the equivalent value of $\Delta a/a$ in the cubic lattice was calculated from the formula

$$\frac{\Delta a}{a} = \frac{(1-\varepsilon)\Delta d}{(1+\varepsilon)d}.$$

Using the above calculated value of Δa we estimate [3] the excess "As" to be around 2.5 x 10^{19} cm^{-3} in this LT GaAs grown at 300C. Low temperature grown GaAs needs to be annealed in situ before growth of the active layer in order to obtain high resistivity GaAs buffer layer. Annealed LT GaAs layer is a non-stoichiometric [3] material containing excess 'As' precipitates and point defects in the matrix of crystalline GaAs layer. The electrical and optical properties of active layers grown directly on such LT buffer layers could be adversely affected due to the fact that excess 'As' could rapidly diffuse into the active layer. Super lattice (SL) layers are known to trap defects and impurities and these have been traditionally employed as buffer layers before the growth of active channel layers. Here, we suggest the introduction of SL buffer on the Low temperature grown GaAs to trap the diffusing excess 'As' since it is known that 'Al' has more affinity for As than Ga. Results of PHEMT structures grown with and without SL buffer on LT GaAs will be discussed later.

2.2 *Pseudomorphic InGaAs layer*

Next we consider the growth of pseudomorphic InGaAs layer in which two dimensional electron gas (2DEG) will be formed. Here the trade-off is among the following parameters. (a) Critical thickness of the strained InGaAs layer [4] (b) mobility and sheet electron concentrating (c) 'In' segregation and re-evaporation [5,6] and (d) AlGaAs/InGaAs interface near which 2DEG transport would take place in the device.

It is known that Indium tends to segregate and Indium incorporation is lower than intended value and it is necessary to adjust the growth conditions to obtain the intended composition in grown layers. We have used Photoluminescence to estimate 'In' composition in InGaAs/GaAs single quantum wells and compare it with PL energy calculated taking into account irregular shape of the QW . We have taken care of the flux transient by providing a ramp of 'In' effusion all temperature.

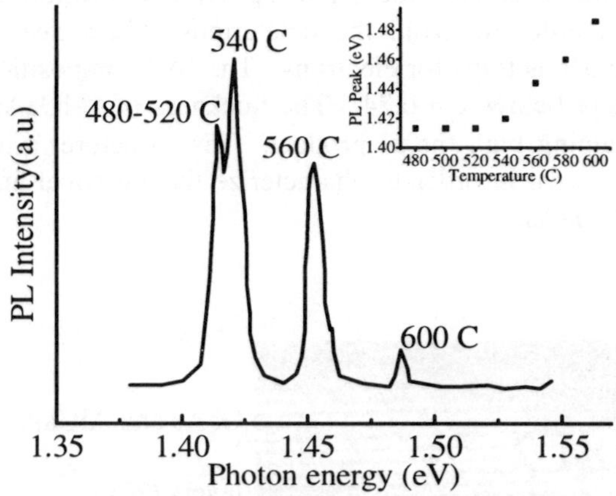

GaAs 500A
$In_{0.1}Ga_{0.9}As$ 75A 600C
GaAs 500A
$In_{0.1}Ga_{0.9}As$ 75A 580
GaAs 500A
$In_{0.1}Ga_{0.9}As$ 75A 560C
GaAs 500A
$In_{0.1}Ga_{0.9}As$ 75A 540C
GaAs 500A
$In_{0.1}Ga_{0.9}As$ 75A 520C
GaAs 500A
$In_{0.1}Ga_{0.9}As$ 75A 480C
GaAs buffer
GaAs Substrate

Figure 3. Photoluminescence spectrum of InGaAs/GaAs MQW structure

We also addressed the problem of 'In' re-evaporation and determined the temperature range in which 'InGaAs' of intended composition can be grown. For this purpose a structure with six 'InGaAs' single quantum well (SQW) of same thickness and nominal intended composition ($x = 0.1$) separated by 500 A GaAs was grown successively by increasing the substrate temperature for each SQW growth. The structure and PL spectrum are shown in Fig. 3.

It can be seen that there are four peaks corresponding to six SQWs. We assign the peaks to different 'In' composition in the Quantum Well(QW) . Higher substrate temperature leads to lower 'In' incorporation in the QW due to re-evaporation from the layer Thus we conclude that for reproducible growth of InGaAs of intended composition, it is necessary to maintain T_s less than 540C.

In a modulation doped [MOD] structure the excess segregated 'In' at the interface could get incorporated during the growth of AlGaAs spacer layer and cause a reduction in the conduction band discontinuity and also roughen the interface. We suggest a method to improve the quality of interface. This involves a growth of a monolayer of GaAs on top of InGaAs layer and subsequently flashing off the excess 'In'. This is expected to result in a 'In' free barrier layer and a smooth interface.

2.3 *Growth of 'AlGaAs' spacer and 'donor' layer*

Many papers[7] discuss the necessity for choosing Al composition in $Al_xGa_{1-x}As$ below x = 0.24 in order to avoid the deleterious effects due to formation of DX centers which act as traps for electrons. The 'Al' composition in the MOD structures was kept below x = 0.24. The thickness of 'AlGaAs' spacer layer is chosen for obtaining high 'ns-μ' product. Test structures with spacer thickness = 65 A were grown in order to characterize the electrical and optical properties of 2DEG in InGaAs.

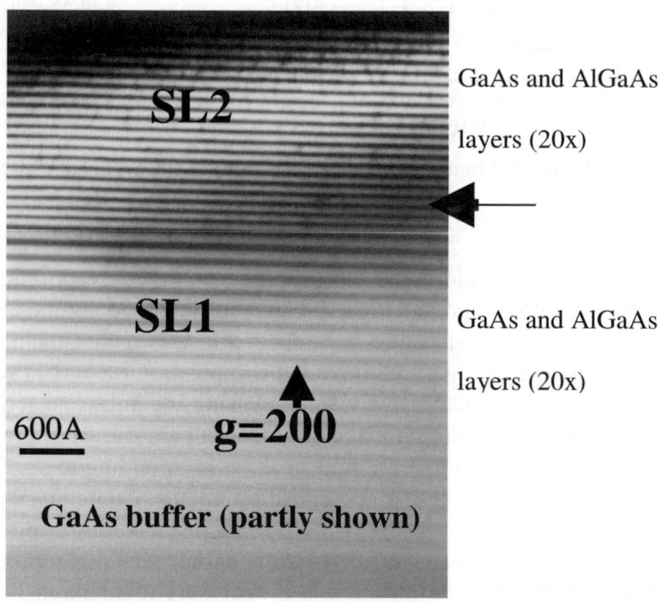

Figure 4. Cross sectional TEM micrograph of a test Superlattice structure containing two superlattices SL1 (GaAs 90A/AlGaAs 115A) and SL2(GaAs 45A/AlGaAs 65A)

Substrate temperature during growth of AlGaAs is an important parameter. It is known that for obtaining best optical properties it is necessary to grow 'AlGaAs' at 700C. However, 'Al' composition in grown layer becomes uncertain due to re-evaporation of 'Ga' which becomes substantial beyond Ts > 680C. Hence, it is necessary to keep the growth temperature below 680C. Increase of T_s before start of spacer layer might roughen the interface and hence the spacer layer was grown around T_s = 620C. In order to evaluate (a) the thickness of thin AlGaAs layers and (b) interface quality we have performed XTEM analysis of a test HEMT structure containing a two period AlGaAs/GaAs Superlattice. The micrograph of the SL portion is shown in Fig 4. The interface quality as seen by HRTEM (not shown) was excellent.

Silicon 'δ' doping concentration was calculated for obtaining 2DEG of around 2 x 10^{12} cm^{-2} in InGaAs and also to provide electrons for the Schottky barrier depletion layer . Substrate temperature was kept at 580C during doping in order to minimize diffusion of 'Si' into the active channel layer.

Although, best quality AlGaAs is grown at substrate temperature 680C, it is necessary to minimize the spread of Si 'δ' doping in the undoped AlGaAs. To obtain high Schottky barrier height, it would be better to keep the AlGaAs undoped and therefore minimize the diffusion of Si towards the surface. In order to achieve this we kept the temperature of growth of AlGaAs at 650 C. Finally, the cap layer of heavily doped GaAs layer was to facilitate Source-Drain ohmic contact formation.

3 Results and Discussion

We show the mobility and sheet carrier concentration obtained in test P-HEMT structures with only SL buffer and also in a structure with a combination of SL and LT buffer in Table 1.

Table 1. Hall measuement Data for the P-HEMT structures

Structure	Mobility (cm^2/vsec)		Sheet Ns(cm^{-2})	
	300K	77K	300K	77K
P-HEMT with only SL buffer	5715	26099	1.5e12	1.7e12
P-HEMT with LT GaAs and SL buffer	4675	19432	2.19e1 2	2.5e12

Fig 5 shows the HRXRD of the P-HEMT structure . In this fig we also show the simulated HRXRD spectra using the simulation program MADMAX[8]. It can be seen that the structure is reasonably sharp and the thickness and composition values in different layers are close to the nominal values.

In fig 6 we show the PL spectra taken at 17K. In this we can identify two peaks viz at 1.605eV and 1.367eV. The peak at 1.605 eV is attributed to the GaAs quantum well in SL buffer. The thickness of the quantum well calculated from the peak position compares well that determined from the TEM measurements. The PL peak at 1.367 eV is from the InGaAs quantum well and is attributed to electron sub band E1-HH1(heavy hole) transition

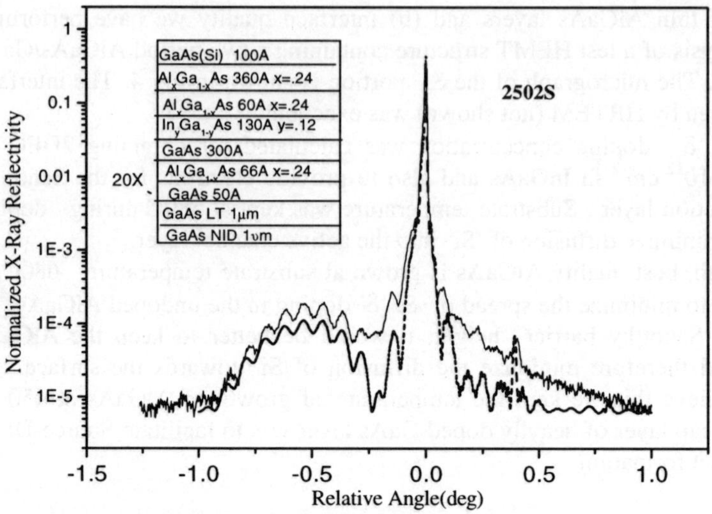

Figure 5. HRXRD (400 reflection)of P-HEMT structure with LT GaAs +SL buffer. Simulated curve is also shown in the figure. The structure is shown in the inset

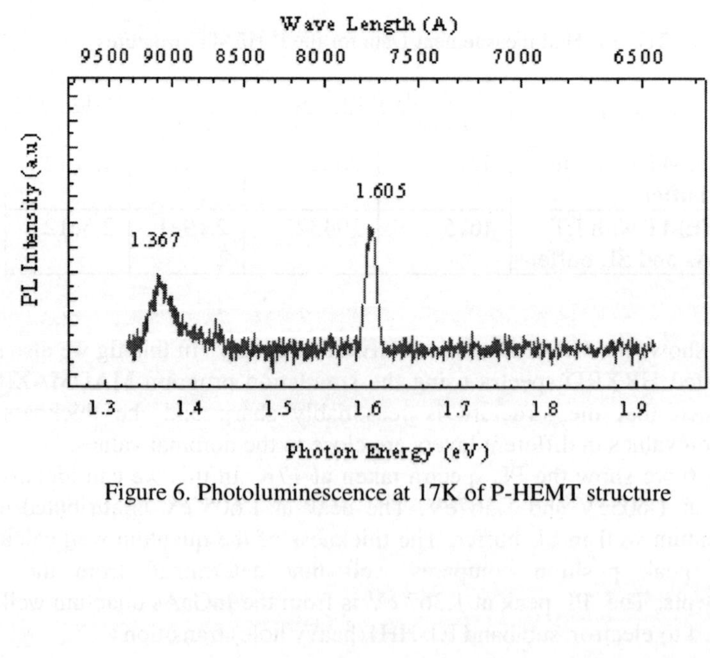

Figure 6. Photoluminescence at 17K of P-HEMT structure

Having optimized the growth condition device structures were grown and 0.5 μm gate length and 100 μm width devices were fabricated which exhibit g_m = 280 ms/mm, a pad capacitance of 0.01 pF and gate to source capacitance Cgs = 15 pF. The cut-off frequency f_t was around 31GHz.

4 Conclusions

Material growth related issues for the MBE growth of AlGaAs/InGaAs/GaAs P-HEMT structures were discussed . Growth of a AlGaAs/GaAs Superlattice on LT GaAs helps in improving the electrical and optical properties of P-HEMT active layer. Results obtained on the P-HEMT structures with LT GaAs containing a SL buffer show very good performance

Acknowledgement

Authors wish to thank Dr S.K Agarwal and Dr V.Kumar for encouragement. We are grateful to Dr H.P.Vyas and his team for providing the device performance results

References

1. Takamiya S,Yoshida N,Hayafuji N,Sonada T,Mitsui S Solidstate Electronics,Vol 38pp1581-1588
2. Smith P.W,Calawa A.R,Chang Lee Chen,.Manfra M.J and.Mahoney L.J, IEEE Electron Device Letters, Vol 9, No 2, Feb 1988, p77-79
3. Liu X, Prasad A., Nishio J, Weber E.R, Lillenthal-Weber Z, Walukiewicz W", Appl.Phys.Lett., 67(2),p279(1995)
4. Helmut Brech, "Optimization of GaAs based High Electron Mobility Transistors by Numerical Simulations " PhD Dissertation, Institute fur MicroElectronik, Technischen Universität Wien,available at www.iue.tuwien.ac.at\publications\PhD Theses
5. Muraki K,Fukatsu S, Shiraki Y,Appl.Phys.Lett,Vol 62, p557(1992)
6. Srinivasan T, Muralidharan R, .Mehta, S.K, .Jain, B P, Singh S.N, Jain,R.K, Kumar V, Vacuum 60(2001)p425-429
7. Mizuta M, Tachikawa M ,Kukimoto H,and Minomura S, Japan. J.Appl.Phys,Vol 24,ppL143-146,1985
8. Computer Simulation software program MADMAX available with Oliver Brandt at Paul Drude Institute , Berlin ,brandt@pdi-berlin.de

SPIN TRANSPORT IN TWO-DIEMSIONAL ELECTRON GAS

T. P. PAREEK AND P. BRUNO
Max Planck Institute fuer Mikrostrukturphysik,
Weinberg 2,
D 06120, Halle, Germany
E-mail: pareek@mpi-halle.de

We persent study of spin transport in two-diemnsional electrong gas in persence of Rashba spin-orbit coupling(RSO). Using Landauer-Büttiker formalism spin resolved conductace of a 2DEG is calculated within singal band tight model in-persence of RSO. It is shown that spin-diffusion length is anisotrpoic, *i..e* it depends on the choice of spin-quantization axis. Further a crossover from 2D to 1D is studied.

1. Introduction

The spin transistor proposed by Datta and Das [1] has attracted a considerabel attention to the problem of spin polarized transport in two-diemnsional elctron gas (2DEG) in persence of the Rashba effect. Datta and Das spin transitor is a hybrid hetrostructures consisting of narrow-gap semicondutor (InAS) connected to feroomagnetic source and drain contacts. It was argued that the conductance of this structure can be modulated by applying an external gate voltage, which can manipulate the precession of conduction elctrons in 2DEG. This requires a filed dependent spin-orbit coupling, which is large and well estabilshed for narrow gap semicondutor heterostructure [4]. Spin-orbit interaction in these structures are governed

For a two dimensional electron gas (2DEG) lying in xy plane the RSO interaction takes the form $H_R{=}\alpha(k_y\sigma_x - k_x\sigma_y) \equiv \vec{B}_R(\vec{k}).\vec{\sigma}$ where $\vec{\sigma} = (\sigma_x, \sigma_y, \sigma_z)$ denotes the Pauli spin matrices,α is RSO coupling parameter and $\vec{B}_R(\vec{k})$ is Rashba field. The direction and magnitude of Rashba field $\vec{B}_R(\vec{k})$ depends on the electron momentum $\vec{k}$. RSO coupling causes a slow spin dephasing by a mechanism known as D'yakonov-Perel'[6], which is a continuous spin precession during electron free flights, contrary to the other spin relaxation processes like Elliot-Yafet[7] and Bir-Aronov-Pikus mechanism [8], which lead to instantaneous spin flip. When an electron propagates, its spin precesses around the direction of $\vec{B}_R(\vec{k})$. The length over which spin precesses by an angle π is known as spin precession length and is related to α as $L_{sp} \approx \frac{\pi}{\alpha}$. Scattering from boundary or impurity changes the direction of Rashba field $\vec{B}_R(\vec{k})$ which depends on the electron wave vector and causes the electron to precess around a new direction. Thus randomizing the precession process causing spin relaxation. The corresponding spin relaxation time τ_ϕ is given by $\tau_\phi \approx 1/(\omega^2\tau_{el})$, where τ_{el} is elastic scattering time [9,10]. The corresponding spin diffusion length is given as, $L_{sd} = \sqrt{(1/2)v_F^2\tau_{el}\tau_\phi} = L_{sp}$ which is independent of the mean free path (L_e). It is important to realize that although the disorder increases the spin-relaxation *time*(motional narrowing effect[9,10]), this increase is exactly compensated by the reduction of the diffusion constant, leading to a spin diffusion *length* that is essentially independent of the disorder. The two latter quantities are probed by different kinds of experiments, namely in a time-resolved or spatially resolved experiment, respectively. We report numerical calculation on a square lattice of length L_x (direction of current flow) and width L_y along the transverse direction, which shows that spin diffusion length is not independent of mean free path(L_e) as predicted by motional narrowing effect[9]. We show that these deviation occurs in two opposite regimes , specifically, (a)In *diffusive* regime ($L_x \gg L_e$, $L_y \gg L_e$) when spin precession length is much larger than the mean free path ($L_{sp} \gg L_e$) (b) In *quasi-ballistic* regime ($L_x \gg L_e$, $L_y \ll L_e$) when spin precession length is much smaller than the mean free path ($L_{sp} \ll L_{el}$). Further it is shown that in case (a) where ($L_{sp} \gg L_{el}$), spin diffusion length can be non-monotonic as a function of Fermi energy due to nonparabolicity of the band. It is also shown that spin relaxation is anisotropic which has important consequences for Datta-Das transistor[1]. These are three new and important result we report in this paper. We would like to stress that the results presented here are obtained within single band tight binding model using recursive green function method [11] which is an exact method and takes into account the *quantum effects* at single particle level.

1.1. *Tight binding Hamiltonian*

The Hamiltonian of a 2DEG lying in xy plane in presence of Rashba spin-orbit coupling, is in continuum given by $H = \hbar^2(k_x^2 + k_y^2)/2m^* + \alpha(k_y\sigma_x - k_x\sigma_y) + V(x,y)$, where $V(x,y)$ is confining potential. For $V(x,y)$ we choose a hard wall confining potential. We discretize the above Hamiltonian on a square lattice of lattice spacing a with N_x sites in the longitudinal direction (current direction) ($L_x=N_xa$) and N_y lattice sites along the transverse direction ($L_y=N_ya$), attached to two ideal non-magnetic leads on left and right. The corresponding tight binding Hamiltonian including RSO [3], takes the form[11,13],

$$\begin{aligned} H = &\sum_{ij,\sigma} \varepsilon c^{\dagger}_{i,j,\sigma} c_{i,j,\sigma} \\ &+ t \sum_{<i,j>\sigma} c^{\dagger}_{i+1,j,\sigma} c_{i,j,\sigma} + c^{\dagger}_{i,j+1,\sigma} c_{i,j,\sigma} + \text{h.c.} \\ &- \lambda_{so} \sum_{i,j,\sigma,\sigma'} \left\{ -c^{\dagger}_{i+1,j,\sigma} c_{i,j,\sigma'} (i\sigma_y)^{\sigma\sigma'} \right. \\ &+ \left. c^{\dagger}_{i,j+1,\sigma} c_{i,j,\sigma'} (i\sigma_x^{\sigma\sigma'}) + h.c. \right\} \end{aligned} \tag{1}$$

here $c^{+}_{i,j,\sigma}$ is the creation operator of an electron with spin σ at site (i,j), ε is a random on-site energy (we use random Anderson model for disorder with width W), t is the hopping energy ($t=\hbar^2/2m^*a^2$, we set t=1 for numerical simulation) and, $\sigma^x_{\sigma\sigma'}, \sigma^y_{\sigma\sigma'}, \sigma^z_{\sigma\sigma'}$ the Pauli matrix elements,summation $<i,j>$ runs over nearest neighbor sites, $\lambda_{so}=(\alpha/2a)$ is the RSO coupling parameter having dimension of energy. The RSO coupling causes spin splitting for $\vec{k}\neq 0$, $\Delta E=2\alpha k$, which is linear in momentum and at the same time causes the spin to precess around the Rashba field $\vec{B}_R(\vec{k})$ with frequency $\omega=\Delta E/2\hbar$.

For later reference we remind that within the Born approximation the mean free path in 2D is given by $L_e=\frac{12\hbar v_f}{2\pi N_{2d}(E_f)W^2}a$, W is strength of Anderson disorder and, $N_{2d}(E_F)$, is density of states. Here we also remind that in 2D the density of states are singular in the middle of the band (this is a general topological property and is not related to any specific model), while goes to a constant near the band edge.

The conductance and spin resolved conductances are calculated using Landauer-Büttiker [14] formalism with the help of non-equilibrium Green's function formalism[11]. The two terminal spin resolved conductance (for a given spin quantization axis) is given by [11]

$$G^{\sigma\sigma'}(\epsilon_F)=\frac{e^2}{h}Tr[\Gamma_1^{\sigma}G_{1N_x}^{\sigma\sigma'+}\Gamma_{N_x}^{\sigma'}G_{N_x1}^{\sigma'\sigma-}] \tag{2}$$

where $\Gamma_{1(N_x)}$ self-energy function for the isolated ideal leads and are given by $\Gamma_{p(q)}=t^2A_{p(q)}$, where $A_{p(q)}$ is the spectral density in the respective lead when it is decoupled from the structure, $G^{\sigma\sigma'+}_{1N_x}$ and $G^{\sigma'\sigma-}_{N_x 1}$ are the retarded and advanced Green's functions of whole structure taking leads into account. The trace is over spatial degrees of freedom. The total conductance is sum of spin-conserved conductance and spin-flip conductance, i.e., $G = G_{sc} + G_{sf}$ where the spin-conserved and spin-flip conductance are $G_{sc} = G^{\uparrow\uparrow} + G^{\downarrow\downarrow}$ and $G_{sf} = G^{\uparrow\downarrow} + G^{\downarrow\uparrow}$ respectively. We point out that in our simulation the injected current is unpolarized since the ideal leads are non-magnetic, however by analyzing the spin resolved conductances we can study the spin relaxation and diffusion phenomena which would be observable when the injector and detector are magnetic. Note that is also possible to obtain such information without magnetic leads, by polarizing and analyzing the electron gas optically.

1.2. *Results and discussion*

Fig. 1 depicts the conductance and spin resolved conductance for different spin quantization axis as a function of length L_x. The width of the channel is fixed and equals L_y=80a. The other parameter are L_e=122a and L_{sp}=104a. This set of parameters corresponds to the case (b) discussed in introduction. The behavior of spin-resolved conductance is different for different quantization axis since the system considered here is confined along transverse y direction the effective Rashba field $\vec{B}_R(\vec{k})$ is almost parallel to y axis hence the spin polarization does not show the oscillation when the spin quantization axis is along y(left panel in Fig. 1) . We note that the spin diffusion length (L_{sd}) is larger than the spin precession length(L_{sp}). Right panel in Fig. 1 shows this clearly where G_{sc} and G_{sf} show oscillatory behavior as a function of length L_x, for lengths larger than the spin precession length (L_{sp}=104a). The period of oscillation is given by L_{sp} and since there are many such oscillation which implies that L_{sd} is larger than the L_{sp} [15]. In a recent model calculation by Silsbee [16] no such oscillation were found. The model used in Ref. [16] ignores spin memory between successive scattering event, while our calculation is exact and hence the spin memory between successive scattering event is taken into account which is a *quantum effect.*

An appropriate quantity which is suitable to study spin diffusion is the polarization of the transmitted electrons, defined as

$$P = \frac{G_{sc} - G_{sf}}{G_{sc} + G_{sf}} \tag{3}$$

From the definition it is clear that the polarization P lies strictly between

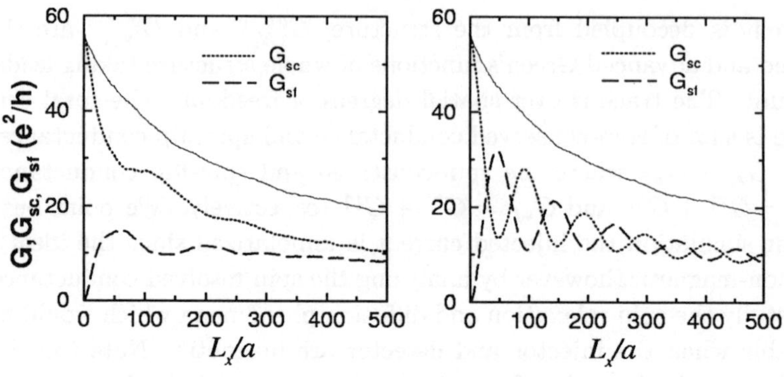

Figure 1. Conductance(solid line) , spin conserved conductance(dashed line) and spin flip conductance(dot-dashed line) as a function of channel length. Spin quantization axis is along y for left panel and along x axis for right panel. The model parameter are $E_f = 1.1t$,$L_y = 80a$, $W = 0.5|t|$, and $\lambda_{so} = 0.03|t|$.

+1 (spin-conserved) and −1 (spin-flip). The polarization corresponding to Fig. 1 is shown in Fig. 2. We see that the polarization is always positive for spin quantization axis y while for spin quantization axis z it shows oscillation of largest amplitude. The amplitude of oscillation is different for different cases signifying that the spin diffusion length is anisotropic [11]. This implies that for the Datta-Das [1] spin-transistor a larger current modulation will be obtained as a function of gate voltage when the magnetization direction of injector-detector ferromagnets are parallel to the z axis [17].

Fig. 3 illustrates the point that in to preserve spin polarization one needs to confine electrons to a width of the order of mean free path and not to tens of mean free path as claimed from real-space Monte Carlo simulation [18]. The mean free path(L_e) for Fig. 3 is $30a$, we see that the polarization for channel width(L_y) $30a$, $50a$, and $80a$ decays much faster as function of channel length compared to the channel width $10a$, and $20a$, which is less than the mean free path. This corresponds to the quasi-ballistic case, *i.e.*,case (b) discussed in the introduction. Also the polarization remains almost unchanged corresponding to channel width $10a$, which is consistent with the 1D limit exhibiting no spin depolarization, since all rotations are along one axis and are commuting [19].

The results presented above were in the quasi ballistic regime corresponding to case (b) in the introduction where $L_{sp} \ll L_e$, so that during free flight electron spin precession was larger than the π. Now we present the result for the regime (a) discussed in the introduction where $L_{sp} \gg L_e$

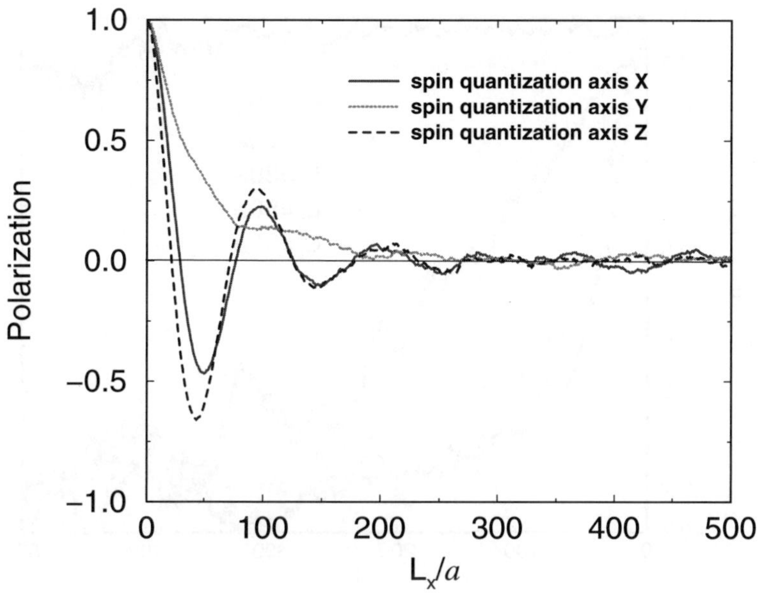

Figure 2. Polarization as function of channel length for spin quantization axis x, y and z.The model parameters are as those for Fig. 1.

so that during free flight electron polarization precesses by an angle smaller than π. The results are shown in Fig. 4, where we have plotted polarization for spin quantization axis y for different mean free paths as a function of spin precession length L_{sp}.

We see in Fig. 4 (left panel) that as we reduces mean free path while keeping spin precession length fixed we observe an enhancement of the polarization P, or in other words polarization decay is reduced, i.e., disorder helps to preserve spin polarization. For $\lambda_{so}/t \geq 0.02$, in the *quasi-ballistic* regime, polarization is enhanced as we increase λ_{so}/t as is seen from the Fig. 4 right panel(curve corresponding to L_e=120a). This corresponds to the case (b) ($L_{sp}/L_e \ll 1$) discussed in introduction, though the condition are opposite to that of case (a) which is motional narrowing regime. Hence in diffusive case spin polarization is enhanced as we increase the dimensional parameter $L_{sp}/L_e \gg 1$ (Fig. 4 left panel) while for quasi ballistic case spin polarization is enhanced when $L_{sp}/L_e \ll 1$(Fig. 4 right panel). In this sense these two opposite regimes behave similar.

Thus encouraged by the results we study the spin-diffusion as function of Fermi energy. A motivation come from the simple observation that the

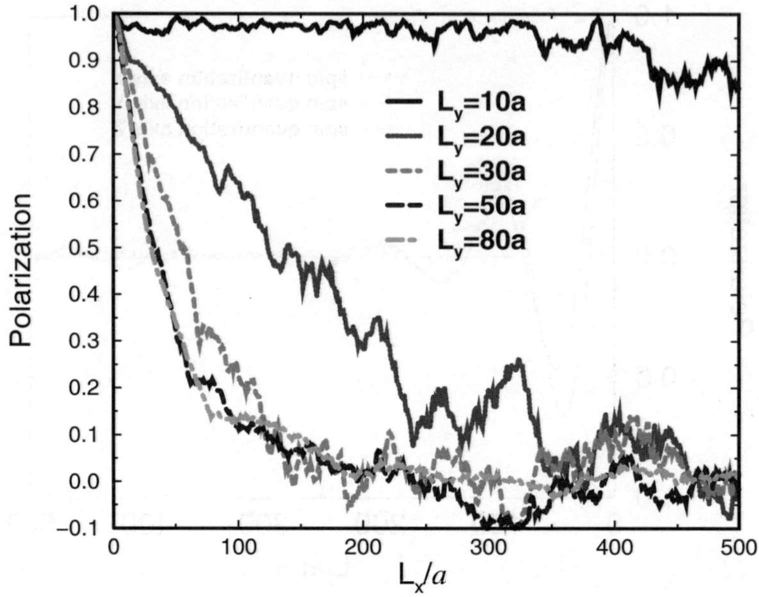

Figure 3. Polarization as function of channel length for different channel width.mean free path and spin precession length are $30a$ and $104a(\lambda_{so} = 0.03|t|)$ respectively. The other parameters are same as those for Fig.1.

mean free path in two dimension behave like $L_e \approx \sqrt(E_f)/(N_{2d}(E_f)W^2)$, any deviation from nonparabolicity in the band will effect the density of states and therefore the spin relaxation. Near the band edges, where energy band can be well approximated by parabola, the mean free path increases as Fermi energy is increased since $N_{2d}(E_f)$ is constant, however as one approaches the band center $N_{2d}(E_f)$ start to diverges logarithmically (Van Hove singularity), this in turn causes mean free path to decrease. This is due to nonparabolicity of the energy band and in recent experiments by Hu *et.al.* [5], it was reported to cause a reduction in RSO coupling λ_{so} by 25%. The reduction in λ_{so} due to nonparabolicity is encouraging since it will increase the spin precession length which can only help to push the parameters in the regime (b) discussed in introduction. From the discussion above we see that as we move away from the band edges, initially the mean free path will increase and then starts to decreases, and will be shortest at the band center. Though presence of disorder will weaken the singularity in density of states, but density of states still remains peaked at the band center as reported recently in Ref. [20]. Hence enhancement of spin

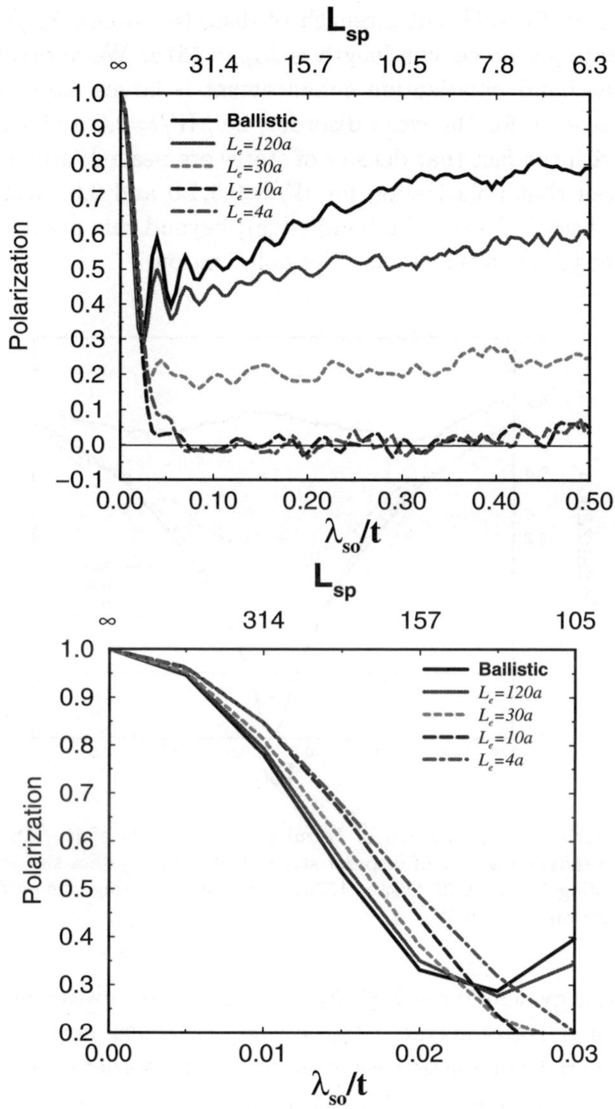

Figure 4. Polarization as function RSO coupling parameter. Different curves corresponds to different mean free path as indicated in the figure. The length L_x and width L_y are both equal to $80a$ and $E_F = 1.1|t|$.

coherence, i.e., polarization should be maximum at the band center. This is clearly seen in Fig. 5, where we have plotted polarization as a function of Fermi energy for different strength of disorder, where $\lambda_{so}/t = 0.02$ or the equivalent spin precession length is $L_{sp} = 157a$. We notice that in the middle of the band polarization enhancement is largest compared to the ballistics case even for the weak disorder, i.e., W/t=0.5 and 1.0, this is in agreement with the fact that density of states are peaked at band center [20]. Also it is seen that polarization for W/t=0.5,1.0 and 2.0, decreases until $|E_f| \approx 2.0$ (measured from the band edge), beyond this point increases in disorder strength always enhances the polarization.

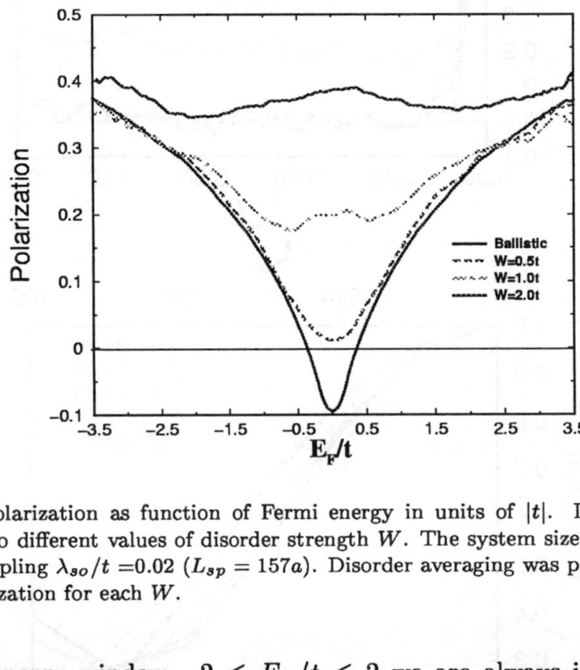

Figure 5. Polarization as function of Fermi energy in units of $|t|$. Different curves corresponds to different values of disorder strength W. The system size is $(80a \times 80a)$. The RSO coupling λ_{so}/t =0.02 ($L_{sp} = 157a$). Disorder averaging was performed for 20 different realization for each W.

In the energy window $-2 \leq E_F/t \leq 2$ we are always in the regime of infinitesimal rotations i.e., $L_{sp}/L_e \gg 1$, hence polarization is enhanced compared to ballistic case irrespective of disorder strength. Beyond $-2 \leq E_f/t \leq 0$ the curve for W/t=2.0 shows an increases in polarization, this is expected since an increase in disorder only helps to decrease the mean free path. This is in agreement with heuristic arguments present above. So from this curve we can safely draw the conclusion that as we move away from band edge, polarization will decrease initially and then will start to increase again till the band center, i.e. polarization shows a non-monotonic behavior as a function of Fermi energy. This is the coupled effect of disorder and nonparabolicity of energy band. This non-monotonic behavior should be seen with reference to recent experiment on n-type GaAs [21], where observed spin life time show a non-monotonic behavior as a function of carrier density. In this experiment carrier density was controlled through doping, which increases the Fermi energy and reduces the mean free path. However the results of Ref. [21] are for 3d bulk material hence we can not make a direct comparison with the experimental result. Another interesting conclusion which can be drawn from Fig. 5 is, in diffusive case spin polarization can be preserved even for wide channels, since disorder helps to preserve spin polarization as soon as nonparabolicity of density of states starts to play role, however weak the disorder may be. This is clearly illustrated in Fig. 5 where all the curves in presence of disorder lies above the curve for ballistic case in the range $-2 \leq E_f \leq 2$. This might have an important implication for the Datta-Das spin transistor [1], since it removes the stringent criterion to confine electrons in one dimension.

2. Symmetry and Anisotropy of conductance of a Ferromagnetic-2DEG interface

In this section we consider the following question, Does the conductance of a FM/2DEG interface depends on the magnetization direction? Búttiker symmetry relation for electrical conduction says that when one flips the magnetization of the Ferromagnet, conductance should remain invariant. However, it does not make any statement about how conductance should behave when one changes the direction of magnetization. Motivated by this , we present the results for the conductance of FM/2DEG interface and point out that the conductance is anisotropic[11]. The results are shown in Fig.6, as is seen clearly when we rotate the magnetization direction in yz plane the conductance changes,and the conductance curve is symmetric with the magnetization angle which is consistent with the Büttiker symmetry relation. This is in contrast to the result of Ref. [12] where they used effective mass Hamiltoniand showed that conductance of a FM/2DEG interface is invariant upon the rotation of magnetization. It was argued in Ref. [12] that the anisotropy which is seen in numerical simulation within tight binding model is a consequence of reduced symmetry of lattice, however we believe that the anisotropy is due to the simultaneous presence of exchange and spin-orbit interaction, though they are spatially separated. The existence of anisotropy is due to the presence of SO interaction which breaks rotational invariance in space.

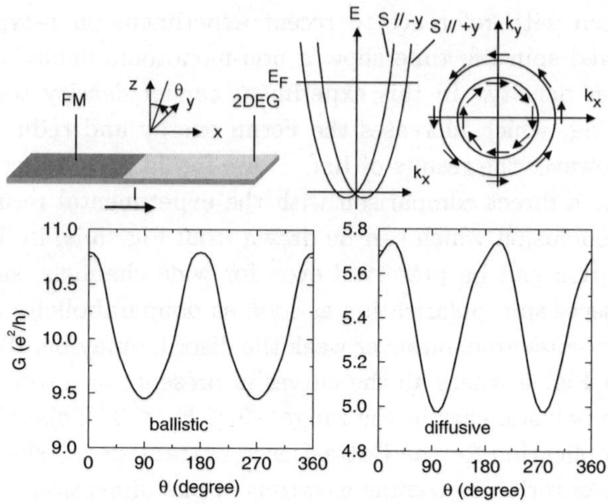

Figure 6. Top: schematic description of the device geometry (left), and of the band-structure (center) and Fermi surface (right) of a 2DEG with Rashba effect; the arrows indicate the direction of the spin-quantization axis. Bottom: Two-terminal conductance of a FM/2DEG junction as a function of the angle θ of the magnetization in yz plane for the ballistic (left) and diffusive (right) cases. Calculations were performed on a 50×50 lattice with $k_F a = 1$, FM exchange splitting Δ given by $\Delta/\varepsilon_F = 0.5$, 2DEG Rashba parameter $k_F\alpha/\varepsilon_F = 0.03$; for the diffusive case (no configuration averaging), the mean-free-path l is given by $k_F l = 30$.

3. Conclusion

In summary we have presented new single band tight binding model including Rashba SO interaction. Using this tight binding models effect of SO interaction on spin and charge transport is studied. Numerical results were presented for spin relaxation in 2DEG in presence of Rashba SO interaction and it was shown contrary to naive expectation that disorder helps to reduce spin relaxation. Also it is pointed out that conductance of a FM/2DEG interface can be anisotropic due to the simultaneous presence of exchange and SO interaction.

Acknowledgments

This work was financially supported by the German Federal Ministry of Research (BMBF).

References

1. S. Datta and B. Das, Appl. Phys. Lett. **56**, 665(1990).

2. G. Prinz, Phys. Today **48**, 58 (1995).
3. Yu. A. Bychkov and E. I. Rashba, Sov. Phys. JETP Lett. **39** 78 (1984).
4. G. Lommer, *et. al.*, Phys. Rev. Lett. **60**, 728 (1988); B. Das *et. al.*, Phys. Rev. B **41** 8278 (1990); J. Nitta *et. al.*, Phys. Rev. Lett. **78**, 1335 (1997); J.P. Heida *et. al.*, Phys.Rev. B. **57**, 11911 (1998); G. Engels, *et. al.*,Phys. Rev. B **55** 1958 (1997); T. Matsuyama *et. al.*, Phys. Rev. B **61**, 15588 (2000).
5. C.-M. Hu *et. al.*,Phys. Rev. B. (1999).
6. M.I. D'yakonov and V.I. Perel', Sov. Phys. JETP **33**, 1053 (1971); Sov. Phys. Solid State **13**, 3023 (1972).
7. R. J. Elliot, Phys. Rev. **96**, 266(1954);Y. Yafet, in *Solid State Physics*, edited by F. Seitz and D. Turnbull (Academic, New York, 1963), Vol. 14.
8. G. L. Bir, A. G. Aronov and G.E. Pikus, Sov. Phys. JETP **42**, 705 (1976).
9. C. P. Slichter, Principles of magnetic resonance, (Harper and Row, New York, 1963);M. Z. Maialle *et. al.*, Phys. Rev. B **47**, 15776, (1993).
10. G.L. Chen et. al., Phys. Rev. B. **47**, 4084 (1993).
11. T.P. Pareek, Phys. Rev. B.**66**, 193301 (2002);T.P. Pareek and P. Bruno Phys. Rev. B.**65**, 241305 (2002)
12. L.W. Molenkamp, G. Schmidt and G.E.W. Bauer, cond-mat/0104109;
13. F. Mireles and G. Kirczenow, Phys. Rev. B **64**, 024426 (2001).
14. M. Büttiker, Phys. Rev. Lett. **57**, 1761 (1986); IBM J. Res. Dev. **32** 317 (1988).
15. A.G. Mal'shukov and K.A. Chao, Phys. Rev. B **61**, R2413 (2000).
16. R.H. Silsbee, Phys. Rev. B **63**, 155305 (2001).
17. H.X. Tang *et. al.* Phys. Rev. B **61**, 4437 (2000).
18. A.A. Kiselev and K.W. Kim, Phys. Rev. B **61**, 13115 (2000).
19. A. Bournel *et. al.* Eur. Phys. J. AP **4** 1 (1998); ibid., Physica B4, 272 (1999).
20. Y. Avishai and Y. Mier, cond-mat/0107591.
21. J.M. Kikkawa *et. al.* Science **277**, 1284 (1997); J. M. Kikkawa and D.D. Awschalom, Phys. Rev. Lett. **80**, 4313 (1998).

STEPPED SILICON TEMPLATES FOR QUANTUM WIRE STRUCTURES

I. K. ROBINSON

Department of Physics, University of Illinois, Urbana, IL 61801, USA

P. A. BENNETT

Department of Physics and Astronomy, Arizona State University, Tempe, AZ 85287, USA

F. J. HIMPSEL

Department of Physics, University of Wisconsin, Madison, WI 53706, USA

During the twenty-year development of synchrotron radiation sources for X-ray diffraction, the application to surfaces and interfaces has been a primary motive for the technological advances. The experimental techniques, which are now well-developed, are described in the introduction of this paper. A recent application has been to understand the quantum physics of one-dimensional electronic structures. Here a vicinal semiconducting surface, Si(557), is used as a template to grow ordered arrays of metallic wires by evaporation of gold. The structure of the resulting Au/Si(557) surface is determined from three-dimensional X-ray diffraction measurements. The data directly mandate a single Au atom per unit cell, which allows the use of a `heavy atom' method in which the Au atom images the rest of the structure. Au is found to substitute for a row of first-layer Si atoms in the middle of the terrace, which then reconstructs by step rebonding and adatoms. This structure is found from theoretical band structure calculations to be consistent with the 1D metallic behavior seen by photoemission.

Surface X-ray Diffraction is a useful experimental method to learn about the atomic-scale structure of surfaces and interfaces. The well-known penetrating property of X-rays inside solid and liquid matter allows us to examine both surfaces and interfaces on an equal footing. Whereas an inherently surface-sensitive method like Scanning Tunnelling Microscopy (STM) can detect the outermost electron shells of the topmost layer, Surface X-ray Diffraction sees the full three-dimensional arrangement of the atom cores. Sometimes, the distinction is important, for example when the electronic density of states is under investigation. In the example presented here, Au/Si(557), the STM is ambiguous about the chemical identity of the prominent features and the electronic states are relevant to its functioning as a quantum wire.

It is commonly not appreciated that Surface X-ray Diffraction obtains its surface sensitivity from *symmetry* considerations alone. Because a surface has lower symmetry than the bulk, its diffraction extends out into regions of reciprocal space where the bulk does not contribute. By choosing to measure in those places, the experiment can be specific about surface structure. Apart from the trivial breaking of the translational symmetry in a reconstructed surface, which leads to entirely new diffraction features, even a "1x1" surface can be determined by Surface X-ray Diffraction because of its Crystal Truncation Rods (CTRs). The Au/Si(557) surface presented here is such an example.

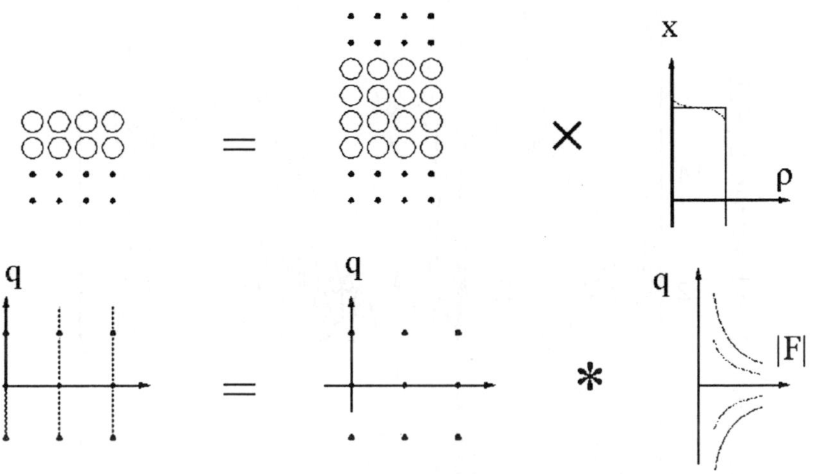

Figure 1. Pictorial view of the origin of Crystal Truncation Rods as a convolution. Top row: surface of a crystal is obtained by ***multiplication*** of an infinite crystal by a step function. Bottom row: after Fourier transformation, the diffraction pattern is obtained by the corresponding ***convolution***.

Crystal Truncation Rods (CTRs) arise because of the abrupt termination of a semi-infinite crystal at a flat surface [1]. They were anticipated by von Laue in his calculation of the contribution of crystallite shapes to powder diffraction profiles [2]. While a mathematical derivation is provided in ref [1], CTRs can be understood graphically to be a convolution, as illustrated in Fig 1. The semi-infinite crystal is represented as a *product* of an infinite crystal with a step function cutting it off in the z-direction. According to the convolution theorem, the diffraction pattern is the *convolution* of the Fourier transforms of these two functions. The Fourier transform of an infinite lattice is a reciprocal lattice of ideal δ-function-like points, given by the usual construction. The Fourier transform of a step function in z is $1/q_Z$, giving the $(1/q_Z)^2$ functional form for the intensity illustrated in Fig 1. Upon convolution, the $(1/q_Z)^2$ functions join together to form continuous lines through reciprocal space, whose intensity diverges at each Bragg point, as shown. In the real situation, absorption (or else dynamical effects) prevents this divergence of intensity. Usually, the surface is a close-packed crystal plane in real space; by the reciprocal lattice construction, the normal to this plane, spanned by q_Z, is an exact direction of reciprocal space, along which the Bragg peaks line up.

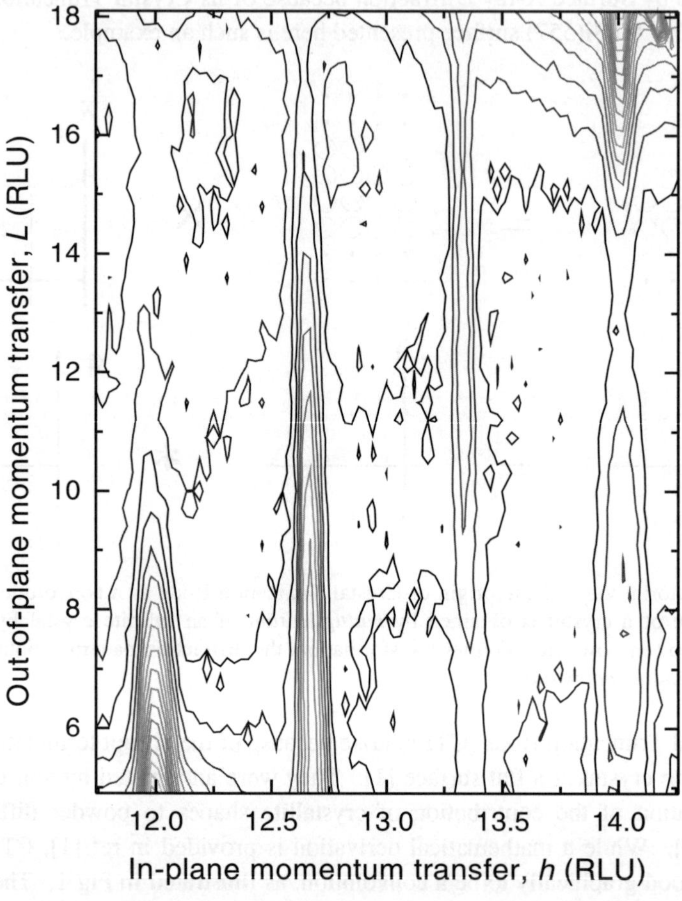

Figure 2. Contour plot showing some of the Crystal Truncation Rods of the clean Si(557) surface. Bragg peaks at positions (12,0,3) and (14,0,20) lie just off the top and bottom edges of the picture. Reconstruction rods due to the 1x3 surface lie in between.

Near to the Bragg peaks, the CTRs are sensitive only to the bulk structure, represented through the usual bulk structure factor. Far away, the CTRs become progressively more surface sensitive, both to the atomic structure and to the degree that roughness is present [1]. The Surface X-ray Diffraction method therefore consists of measuring the intensities of a large number of CTRs and comparing with the surface structure factor calculated as the amplitude sum of the surface and bulk. A widely used computer program, called ROD, allows the refinement of the atomic coordinates in three dimensions [3]. ROD is not useful for *ab initio* modelling because of the phase problem inherent to all X-ray Diffraction methods; a close starting model is required. Because of its simplicity, having a single heavy atom (Au) per unit cell, the Au/Si(557) structure "solves itself". As first shown for small molecules [4], the pair correlation (Patterson) function, obtained by Fourier transformation of the measured intensities, closely resembles the electron density of the structure combined with its inverted "twin" image.

The clean Si(557) surface was prepared on a wafer fragment by heating it in UHV to 1200°C. As previously reported [5], this resulted in a 3x1 reconstruction, believed to be comprised of triple bunches of steps. Part of its diffraction pattern is shown in Fig 2. The indexing of the reciprocal lattice appears unusual, but is required to meet the necessary condition of placing the q_z-direction along the (557) surface normal. The 11$\underline{1}$ bulk reflection indexes as (12,0,3), while the 220 bulk reflection indexes as (14,0,20) in these "surface" coordinates [6]. The (12,0,L) and (14,0,L) CTRs can be clearly seen as ridges of intensity in the contour plot, rising strongly towards the two Bragg peaks, just off the edge of the frame. Two additional rods of intensity lie in between the CTRs at 1/3-order positions, which are due to the reconstruction.

This (presumably) step-bunched substrate was then used as a template to prepare the Au/Si(557) surface by evaporation of 0.2 monolayers of Au onto the substrate at 600°C then annealing briefly at 900°C. The annealing permits the substrate atoms to rearrange, resulting in a "1x1" surface, containing only the (12,0,L) and (14,0,L) CTRs and no 1/3-order rods [6]. This is the structure we solved by measurement of the intensities of the CTRs [6]. The Patterson function gave us directly the termination of the Si lattice at the surface. Some additional peaks in the Patterson suggested small modifications, which were tested and refined using ROD to give the final structure of Fig 3.

In describing the final structure, the most noteworthy feature is that the Au atom substitutes for a row of Si atoms right in the *middle* of the terrace. The substitutional behavior is new in the sense that all previous structures of Au on Si surfaces have been ambiguous about the binding site(s). Here it is very clear that almost ideal substitution takes place in the upper bilayer, resulting in Au-Si bondlengths of 2.46Å and 2.35Å, which are within the limits of error from Si-Si at 2.35Å; the Au appears to substitute in the Si lattice with relatively little strain. The second main feature is that the step edge reconstructs to form a 5-membered ring, but removes a dangling bond. The third feature is a row of adatoms attached to (otherwise) dangling bonds, as shown. The adatoms were only just detectable with the current number and quality of data (349 independent measurements, 47 structural parameters), reducing the chisquare from 7.5 to 7.1 for the final model.

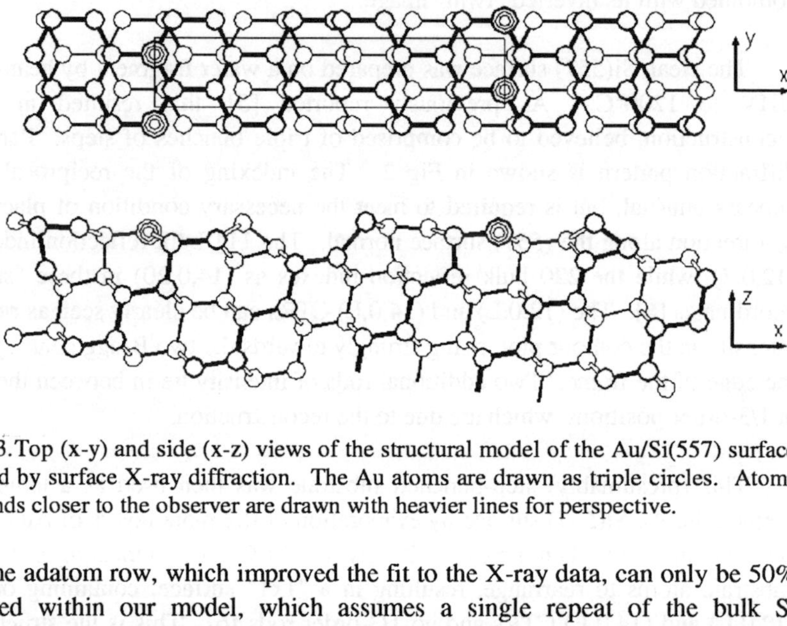

Figure 3. Top (x-y) and side (x-z) views of the structural model of the Au/Si(557) surface obtained by surface X-ray diffraction. The Au atoms are drawn as triple circles. Atoms and bonds closer to the observer are drawn with heavier lines for perspective.

The adatom row, which improved the fit to the X-ray data, can only be 50% occupied within our model, which assumes a single repeat of the bulk Si structure in the y-direction (see Fig 3). This is because it requires three upper bilayer dangling bonds on a Si(111) terrace to attach an adatom, so two unit cells along y are needed. In fitting the data, a 50% occupancy per single unit cell was therefore assumed. The finding of reduced occupancy might account for the doubled periodicity of one of the two rows seen per unit cell seen in the STM results, an example of which is compared with our model in Fig 4.

The main purpose of Fig 4 is to compare the relative distances along the x-direction between the protruding features of the model and STM image. The row spacing of the STM is unfortunately compatible with two possible spacings of the atomic model, so the identification remains ambiguous: the second row could be either the Au or the step edge. We identify this other row to be the step edge for two reasons. Firstly, the peaks of both rows appear alike in STM, even at different tunnelling voltages. Secondly, even though the step edge is not expected from our model to have a doubled period, it shows one nevertheless. Since the adatom is known to induce considerable strain parallel to the surface, it is conceivable that a buckling of the step-edge atoms could result. Such a buckling was already found to lower the surface energy in an *ab initio* calculation [7]. Conversely, buckling of an Au row is not supported. This prediction, which would explain the STM findings, is testable in a future X-ray diffraction experiment, if the surface can be ordered sufficiently to obtain half-order diffraction peaks along the y-direction.

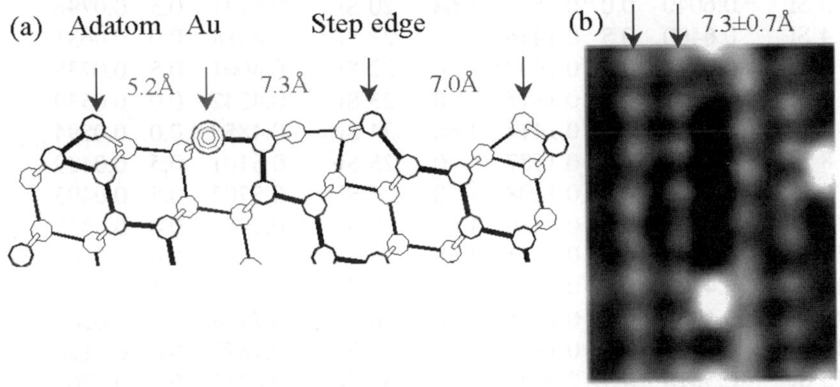

Figure 4. (a) Side (x-z) view of the refined structure of Au/Si(557) showing the spacings between the three most prominent features. An atom numbering scheme is introduced. (b) STM image of Au/Si(557) showing the separation of the two rows of bright spots. The image represents a top view of the surface indicated by the x-y directions shown. The bright spots are identified here as Silicon atoms. The image was obtained using methods described in ref [5].

Refined atomic coordinates are provided in Table 1. Crystallographic notation is used, giving coordinates as fractions of an orthorhombic unit cell that contains two copies of each atom. The numbering scheme is marked using labels in Fig 5. Fig 5 also denotes the bond lengths that are identified in the model. Excluding the adatom, these were found to be distributed with an average of 2.37Å and a standard deviation of 0.07Å. The bulk Si bond length is 2.35Å. The adatom geometry showed a back-bond length of 2.11 ± 0.2Å, although this is a little unreliable because of its low occupancy. There is a significant trend in bond angles that affects the overall appearance of the surface in Fig 5. Close to the underside of the step, the geometry is close to tetrahedral near 109°, suggesting sp^3 bonding. This tends towards trigonal near 120°, indicating sp^2 bonding, at the outermost end of the terrace.

Table 1. Refined coordinates of the atomic positions using the numbering scheme of Fig 5. The coordinates are fractions of the orthorhombic unit cell constants, a=38.208Å, b=3.840Å and c=54.035Å. Two copies of each atom exist in the centered orthorhombic unit cell, related by the translation vector (0.5,0.5,0). The column denoted B is the Debye-Waller factor in $Å^2$.

Atom	x	y	z	B	Atom	x	y	z	B
1 Si	0.5134	0.5	0.1598	3.15	18 Si	0.7291	0.0	0.0922	0
2 Si	0.5455	0.0	0.1544	3.15	19 Si	0.7529	0.5	0.0751	0
3 Si	0.6040	0.0	0.1533	1.68	20 Si	0.8131	0.5	0.0796	0
4 Si	0.6400	0.5	0.1456	0	21 Si	0.8384	0.0	0.0631	0
5 Au	0.7042	0.5	0.1493	0	22 Si	0.4001	0.5	0.0735	0
6 Si	0.7355	0.0	0.1375	0	23 Si	0.4242	0.0	0.0530	0
7 Si	0.8000	0.0	0.1414	1.68	24 Si	0.4853	0.0	0.0594	0
8 Si	0.8383	0.5	0.1587	0	25 Si	0.5101	0.5	0.0429	0
9 Si	0.8255	0.5	0.1208	0	26 Si	0.5707	0.5	0.0493	0
10 Si	0.8853	0.5	0.1330	1.68	27 Si	0.5960	0.0	0.0328	0
11 Si	0.4123	0.5	0.1171	1.68	28 Si	0.6566	0.0	0.0404	0
12 Si	0.4728	0.5	0.1237	1.68	29 Si	0.6818	0.5	0.0227	0
13 Si	0.4964	0.0	0.1025	0	30 Si	0.7424	0.5	0.0303	0
14 Si	0.5572	0.0	0.1114	0	31 Si	0.7677	0.0	0.0126	0
15 Si	0.5816	0.5	0.0926	0	32 Si	0.8283	0.0	0.0202	0
16 Si	0.6405	0.5	0.1038	0	33 Si	0.8535	0.5	0.0025	0
17 Si	0.6667	0.0	0.0854	0	34 Si	0.4141	0.0	0.0101	0

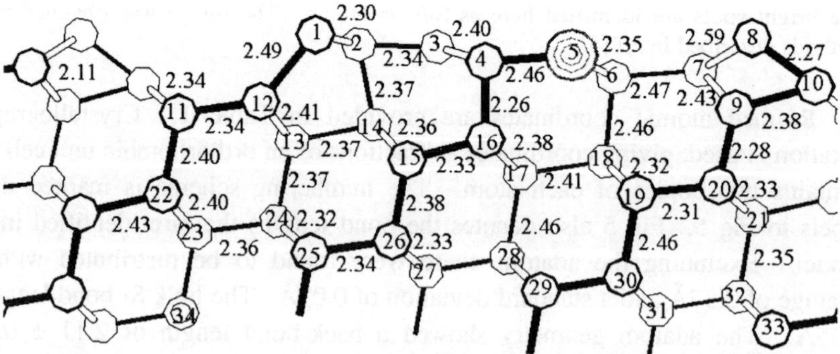

Figure 5. Side view of the refined structure of Au/Si(557) showing the numbering scheme and the bondlengths (in Å) between the atoms which were refined.

Ab initio calculations for the structure of Fig 2 have already been completed [7]. The calculation compared a number of possible configurations and independently found the same step reconstruction [7]. The adatom was found to be acceptable to the total energy, but without a strong difference. The middle Au substitutional site was also confirmed to be favorable and found to contribute to strongly hybridized Au-Si electron bands which cross the Fermi level. These are the states that contribute to the one-dimensional conductivity in this system.

The X16A beamline facility where the measurements were made at NSLS is supported under DOE contracts DEAC02-98CH10886 and DEFG02-91ER45439. This research was supported by the National Science Foundation under contracts DMR 98-76610, DMR 99-81779, DMR 96-32635 and DMR 98-15416.

References

1. I. K. Robinson, Phys. Rev. B **33** 3830 (1986)
2. M. von Laue, Annalen der Physik [5] **26** 55 (1936)
3. E. Vlieg, J. Appl. Cryst. **33** 401 (2000)
4. H. B. Dyer, Acta Cryst. **4** 42 (1951)
5. A. Kirakosian, R. Bennewitz, J. N. Crain, T. Fauster, J. L. Lin, D. Y. Petrovykh and F. J. Himpsel, Appl. Phys. Lett. **79** 1608 (2001)
6. I. K. Robinson, P. A. Bennett and F. J. Himpsel, Phys. Rev. Lett. **88** 096104 (2002)
7. D. Sanchez-Portal, J. D. Gale, A. Garcia and R. M. Martin, Phys. Rev. B **65**, 081401 (2002)

SCANNING TUNNELING MICROSCOPY STUDY OF EPITAXIAL GROWTH OF SI AND GE ON SILICON DURING GROWTH

BERT VOIGTLÄNDER

Institut für Schichten und Grenzflächen, Forschungszentrum Jülich, 52425 Jülich, Germany

E-mail: B.Voigtlaender@fz-juelich.de

We use a scanning tunneling microscope (STM) capable of imaging the growing layer during MBE-growth at high temperatures. This method (MBSTM) opens the possibility to follow MBE growth processes dynamically on the atomic scale and gives access to the evolution of specific features during growth. The influence of surface reconstructions on growth kinetics can be studied directly. For the case of growth of Si islands on Si(111) we find lateral growth of rows of the width of the 7×7 reconstruction unit cell at the edges of twodimensional islands. This leads to a kinetic stabilization of magic island sizes. The evolution of size and shape of individual {105} faceted Ge islands (hut cluster) on Si(001) is measured during growth. A slower growth rate is observed when an island grows to larger sizes. This behavior can be explained by kinetically self-limiting growth.

1. Introduction

Recently scanning tunneling microscopy (STM) has become a powerful method to study epitaxial growth on the atomic level. However, STM has been limited in the past to "snapshots" of certain growth stages. The growth was interrupted at a specific coverage and the sample was quenched to room temperature and transferred to the STM for imaging. The main disadvantage of this mode of STM operation is that local growth structures which are accessible by a real space microscopy cannot be studied in their dynamic evolution during growth. Recently some effort was made to overcome these disadvantages [1,2,3].

Here we describe a mode of STM operation which opens the possibility to observe the dynamics during epitaxial growth "in vivo" (MBSTM). The experiments were performed in a standard ultra-high vacuum system. Due to the open design of the STM the molecular beam from an evaporator can be directed towards the sample while the STM is scanning the surface [4,5].

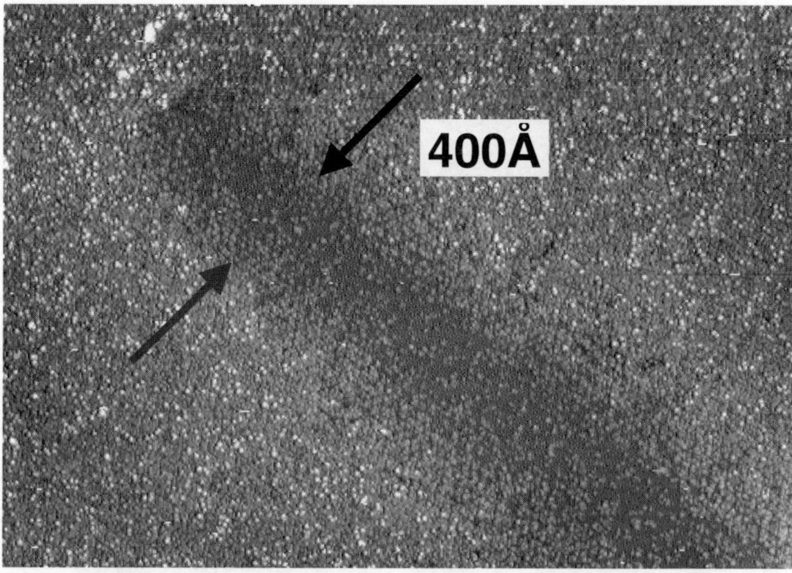

Figure 1. gray-scale image of the silhouette of the STM tip on the surface produced by the molecular beam. The Ge covered region is shown brighter while the darker region was shaded by the STM tip. The tip diameter can be measured to be 400 Å.

During imaging the sample is held at MBE growth temperatures of 600-900 K.

Using this method of simultaneous scanning and molecular beam epitaxy the area behind the tip is shaded. This area can be measured if we evaporate Germanium at room temperature (small diffusion length) while the tip is fixed (i.e. not scanning). Scanning this area afterwards shows the silhouette of the tip on the surface (fig. 1). The tip diameter can be measured to about 400 Å so that for images of several thousand Å scan range the shaded area is sufficiently small.

2. Influence of the (7×7) reconstruction in Si/Si(111) epitaxy

During the epitaxial growth of 2D Si islands on Si(111) the growth of a selected island can be observed as a function of time. Figure 2(a-f) shows STM images from a growth sequence of such an island. In (a), the shape of the island is triangular. Images (b-d) show the same island at a later stage during growth. As shown by images (b-d), growth proceeds by advancement of a row of a certain width along the right island edge. The position of

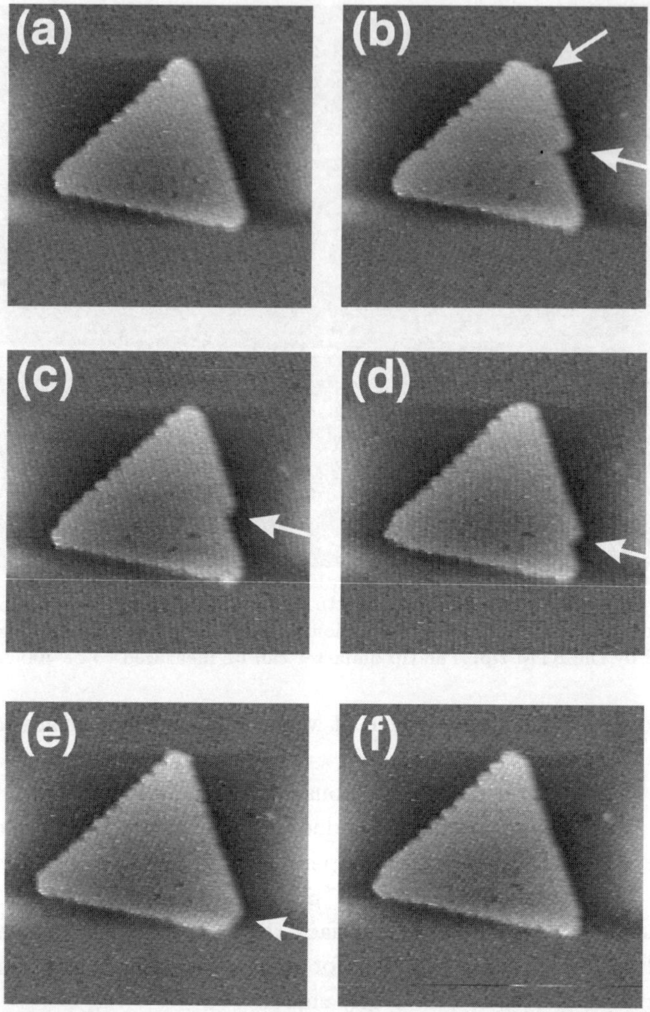

Figure 2. Sequence of images showing the lateral growth of a triangular Si(111) island. A row of the width of the 7×7 unit cell is growing along the right edge of the island (b-f). The image size is 500×500 Å^2, T=575 K. The complete growth sequence is available as a movie on the World Wide Web: http://www.fz-juelich.de/video/voigtlaender.

the kink at which the row is ending is shown by an arrow in (b-d). An analysis of the width of this row and further atomically resolved images show that the width of such a row is 27 Å, which is just the width of one 7×7 reconstruction unit cell [6].

In the following we discuss the influence of the 7×7 reconstruction on

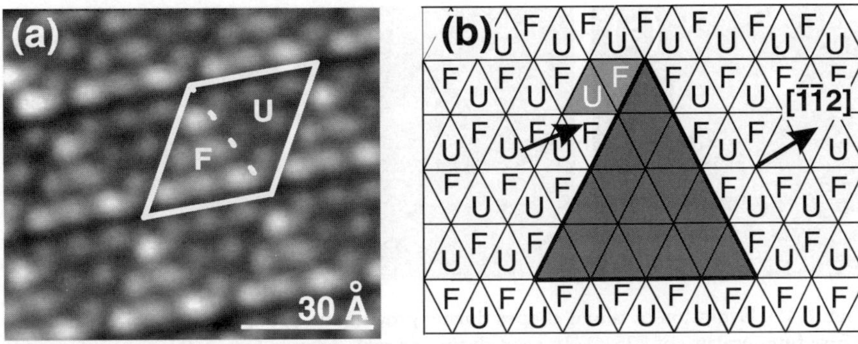

Figure 3. (a) STM image of the Si(111) - 7×7 reconstruction. The white protrusions are the Si adatoms. The unit cell is indicated by a white rhombus. A dashed line divides two triangular subunits. Since the image is taken at negative sample bias (-2 V), the adatoms in the faulted half of the unit cell (F) appear brighter than those in the unfaulted half (U). (b) Arrangement of the U and F parts of the 7×7 unit cell on the substrate around a triangular island (shown in gray).

the growth behavior. The rhombic unit cell of this reconstruction consists of two triangles. One of this triangles has a stacking fault in the surface layers, relative to the substrate stacking (F-HUC). The other triangle of the reconstruction unit cell is unfaulted relative to the substrate (U-HUC), fig. 3(a).

During lateral growth of an island, the surface reconstruction of the substrate has to be lifted and the substrate atoms have to rearrange to the bulk structure. This transformation of the reconstructed surface layer towards the bulk structure is a general phenomenon which has to occur in any epitaxial growth at reconstructed surfaces. It requires a higher energy to lift the reconstruction of the F-triangle as compared to the U-triangle[7]. To lift the reconstruction in the U-triangle, only atoms in the uppermost adatom layer have to rearrange. This is associated with a relatively low energy barrier. Lifting the reconstruction of the F-triangle requires the removal of the stacking fault in the layer below the adatoms. This rearrangement of atoms in deeper layers is associated with a larger energy barrier. This should lead to a high activation barrier for overgrowth of the F-triangle compared to overgrowth of the U-triangle.

In Fig. 3(b), a Si island (gray) and the U and F triangles of HUCs of the surrounding reconstructed substrate surface are shown. Due to the crystallographic orientation of the island, it is surrounded only by substrate F-HUCs. This means that further lateral growth (initiated by the

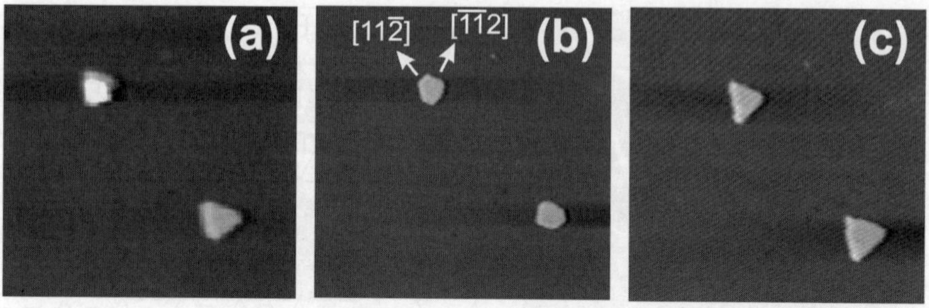

Figure 4. Evolution of island shapes during growth and growth interruption (at a substrate temperature of 725 K). Imaged during growth, islands have a triangular form (a). After 18 minutes of annealing at 775 K, the islands have changed their shape to hexagonal (rounded) (b). After growth commenced again, the island shapes changed back to triangular (c). In all three images, snapshots of the same two islands are shown. The image size is 550 Å × 550 Å.

overgrowth of an F-triangle) is hindered by a high energy barrier. Once an F-triangle has nucleated, the neighboring U-triangles can be overgrown more easily (no stacking fault has to be removed). The overgrowth of the next F triangle is facilitated by the existence of a "macro kink" (arrow in Fig. 3(b)). Here the cost of the stacking fault energy is reduced by a gain in the island edge energy: The edge length is reduced after growth of an F triangle. Therefore, neighboring U and F units can be overgrown in quick succession, leading to the fast growth of a stripe of the width of the 7×7 unit cell.

Due to the quick growth of a row once it is nucleated and a longer time until the nucleation of the next row the island shape of a completed triangle is found often in snapshot images of the island morphology. Islands of complete triangular shape are kinetically stabilized. This higher stability for the closed shell triangular islands leads to pronounced peaks of islands of magic sizes in the measured island size distributions [8].

The two-dimensional triangular islands form during submonolayer growth of Si on Si(111) are 1 BL high (1 BL = one (111)-bilayer = 3.1Å). In the following it will be shown that this is a *kinetically* limited growth shape evolving during growth. Equilibration of these islands without external flux results in a transition to the hexagonal equilibrium form

In Fig. 4 two islands are imaged first during growth (Fig. 4(a)), then after a growth interruption of 18 min (Fig. 4(b)), and finally after the growth was resumed (Fig. 4(c)). These experiments show that when growth is interrupted, the islands quickly lose their triangular shapes: First, the atoms from the apex regions detach creating islands with rounded corners. Subsequently, the island shapes more slowly turn into a hexagon-like (compact) shape as shown in Fig. 4(b). When growth is continued, the islands quickly resume triangular shapes again. This evolution of growth morphology from triangular islands during evaporation to hexagonal islands upon equilibration of the surface at 725 K and the final transition back to triangular islands when the external flux is resumed and the equilibrium conditions are no longer maintained is clearly shown in Fig. 4. This process shows that the triangular island shapes are non-equilibrium ones while equilibrium shapes are close to hexagons.

The equilibrium shape of an island is the form for which the energy of the island is minimized. If the steps in the $[\bar{1}\bar{1}2]$ and the $[11\bar{2}]$ direction were equivalent, the equilibrium shape would be an equilateral hexagon. During annealing the form of the islands approaches such a hexagon. However several observations after long times of annealing show that the edges perpendicular to $[11\bar{2}]$ are a bit shorter than the edges perpendicular to the $[\bar{1}\bar{1}2]$ direction. This indicates that the edges perpendicular to $[11\bar{2}]$ have a higher energy.

3. Kinetically self-limiting growth of Ge islands on Si(001)

In Ge epitaxy on Si(001) three-dimensional Ge islands with $\{105\}$ facets nucleate after the completion of the 2D wetting layer [9]. The island edges are oriented along $< 100 >$ directions, i.e. at 45° to the direction of the reconstruction dimer rows. The strain of these islands is partially relaxed elastically, but they are coherent with the substrate lattice (i.e., free of dislocations between the substrate and the hut clusters) [9]. The evolution of the growth of these hut clusters is observed by STM during growth. A sequence of STM images for increasing coverage is shown in fig. 5. From the images for instance the volume of the hut clusters can be calculated.

Figure 6 shows the evolution of the volume of several individual islands as a function of total deposited coverage. The initially higher growth rate of individual islands just after the nucleation, indicated by initially large slopes in fig. 6, decreases when the islands grow to a larger size. In the following, we use a model which shows that kinetically self-limiting growth explains the observed slower growth for larger island sizes. This kinetic model for the growth of hut clusters relies on a barrier for the nucleation of each successive atomic layer on the $\{105\}$ facets [10,11]. The barrier for the nucleation of a new facet increases with the facet size. This size de-

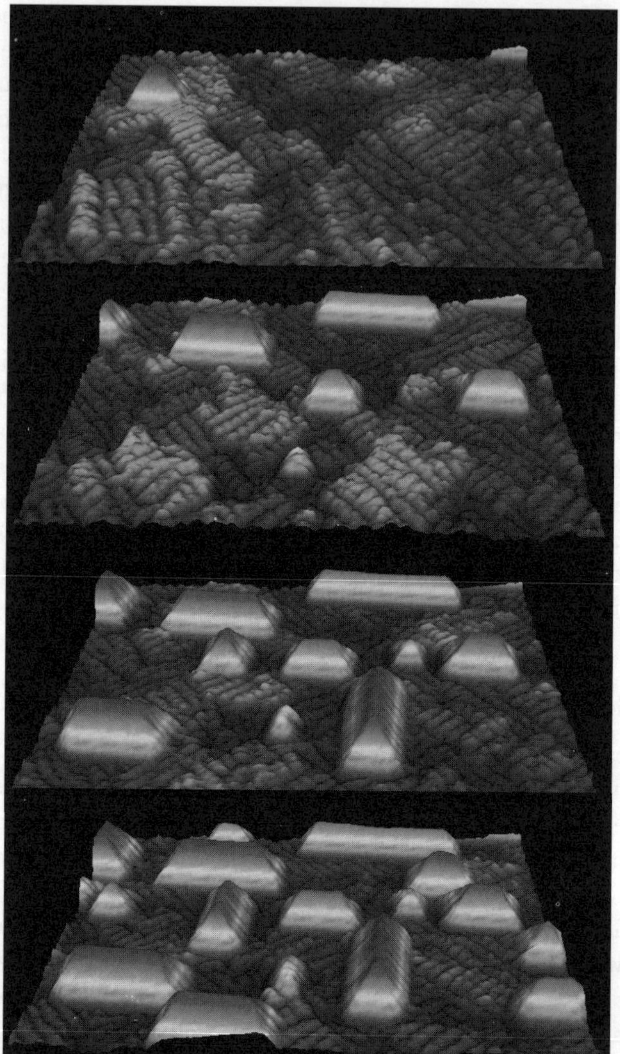

Figure 5. Perspective view of STM sample images of the hut cluster growth as function of coverage beyond the wetting layer. (0.07 ML, 0.46 ML, 1.33 ML and 1.81 ML from the top to the bottom; Image area: 1300×1000 Å^2, T=575 K. The complete growth sequence is available as a movie on the World Wide Web: http://www.fz-juelich.de/video/voigtlaender/.

pendent nucleation barrier for the repeated overgrowth of the {105} facets is the reason for the self-limiting growth behavior [11]. Results of the model

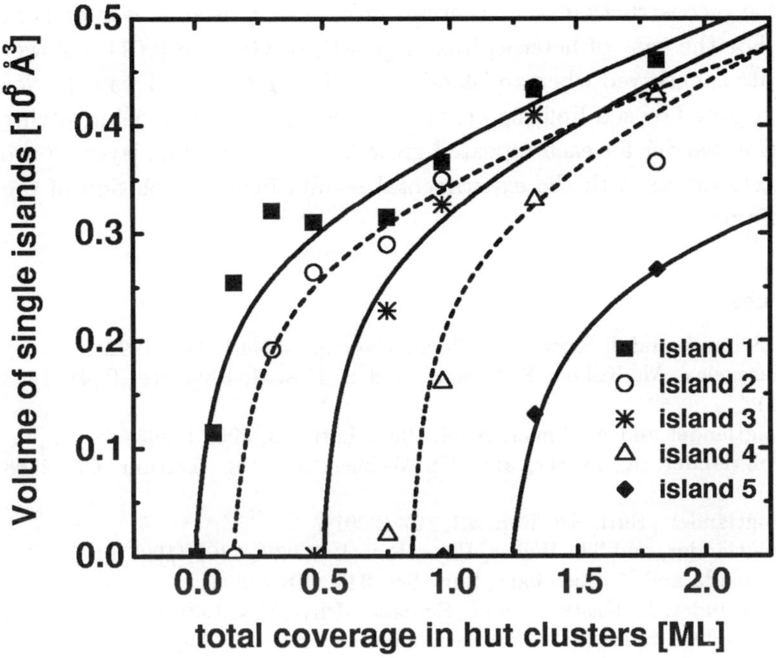

Figure 6. Evolution of the volume of individual hut clusters. The different symbols correspond to different individual islands. The size evolution shows self-limiting behavior: The initially larger growth rate (large slope) just after the nucleation decreases when the islands grow larger. Results of a model calculation of kinetically self-limiting growth including a kinetic energy barrier for the nucleation of new material on the facets are shown as solid and dashed lines.

calculations for the evolution of the hut cluster volume are shown as solid and dashed lines in fig. 6 and are in good agreement with the experimental data. The experimentally observed slower growth rate for larger islands is clearly reproduced. This indicates, that a kinetic self-limitation is effective during the growth of larger hut clusters. This kinetic self-limitation arises due to an energy barrier for the nucleation of new material on completely filled {105} facets.

4. Summary

In summary we have shown that STM can be operated in real time during molecular beam epitaxy at high temperatures. MBSTM opens the possibility to observe the growth history of single growth features in real time. In the case of the homoepitaxial growth of Si on Si(111) the influence of the surface reconstruction on the lateral growth of Si islands was demonstrated. For the case of heteroepitaxial growth of Ge on Si(001) a slower growth rate is observed when an island grows to larger sizes. This behavior can be explained by self-limiting growth. A kinetic growth model involving a nucleation barrier for each repeated growth of a new atomic layer on the {105} facets agrees with the experimental results for the evolution of the island volume.

References

1. A.K. Schmid, and J. Kirschner, Ultramicroscopy **42-44**, 483 (1992).
2. T. Hasegawa, M. Kohno, S. Hosaka, and S. Hosoki, Phys Rev.B **48** 1943 (1993).
3. B. Voigtländer and A. Zinner, Appl. Phys. Lett. **63**, 3055 (1993).
4. B. Voigtländer, A. Zinner, and Th. Weber, Rev. Sci. Instrum. **67**, 2568 (1996).
5. B. Voigtländer , Surf. Sci. Rep. **43**, 127 (2001).
6. B. Voigtländer, and Th. Weber, Phys. Rev. Lett. **77**, 3861 (1996).
7. W. Shimada and H. Tochihara, Surf. Sci. **311**, 107 (1994).
8. B. Voigtländer, M. Kästner, and P. Smilauer, Phys. Rev. Lett. **81**, 858 (1998).
9. Y.W. Mo, D.E. Savage, B.S. Schwartzentruber, and M.G. Lagally, Phys. Rev. Lett. **65**, 1020 (1990).
10. K.M. Chen, D.E. Jesson, S.J. Pennycook, T. Thundat, and R.J. Warmack, Mat. Res. Soc. Symp. **399**, 271 (1996).
11. M. Kästner and B. Voigtländer, Phys. Rev. Lett. **82**, 2745 (1999).

GROWTH OF SELF-ASSEMBLED EPITAXIAL GERMANIUM NANOISLANDS ON SILICON SURFACES BY MOLECULAR BEAM EPITAXY

D. K. GOSWAMI, B. SATPATI, P. V. SATYAM AND B. N. DEV

Institute of Physics, Sachivalaya Marg, Bhubaneswar-751005, India

Self-assembled epitaxial Ge nanoislands have been grown on Si(100) and Si(111) surfaces by molecular beam epitaxy in a compact custom-made MBE system. Reflection high energy electron diffraction (RHEED) studies during growth shows layer to island growth transition as expected for Stranski-Krastanov growth. The samples have been characterized by *in-situ* scanning tunneling microscopy (STM) and *ex-situ* transmission electron microscopy (TEM). Growth under identical conditions appears to form larger islands on Si(110) surfaces compared to Si(100).

1. Introduction

Epitaxial layer growth and layer morphology are governed by the interplay between surface free energies and elastic energy. In the case of heteroepitaxial growth there are three different growth modes:[1] (a) Frank-van der Merwe (FM) or layer-by-layer growth, (b) Volmer-Weber (VW) or island growth and (c) Stranski-Krastanov (SK) or layer-plus-island growth. Which growth mode will be adopted in a given system depends on the surface free energy of the substrate, σ_s, that of the film, σ_f, and the interfacial energy σ_i. Layer-by-layer growth mode occurs when $\Delta\sigma = \sigma_f + \sigma_i - \sigma_s \leq 0$. The condition of FM mode growth is rigorously fulfilled only for homoepitaxy (same film material as that of the substrate) where $\sigma_s = \sigma_f$ and $\sigma_i = 0$. If the FM mode growth condition is not fulfilled then three dimensional crystals form immediately on the substrate (VW mode). For a system with $\Delta\sigma \leq 0$ but with a large lattice mismatch between the substrate and the film, initial growth is layer-by-layer. However, the film grows as a strained layer. As the film grows in thickness it stores an increasingly large strain energy. This strained epilayer system can lower its total energy by forming isolated thick islands in which the strain is relaxed by interfacial misfit dislocations, leading to SK mode growth in these strained systems.

SK growth mode occurs when there is a lattice mismatch between the substrate and the epilayer causing the epilayer to be strained. Growth of dot-like self-assembled islands is a consequence of that. Wire-like islands can grow from dot-like islands via a shape transition which helps strain relaxation. This shape transition has been predicted[2] and observed[3]. The dot islands grow in a shape following the symmetry of the substrate. For example, equilateral triangular[3] and square islands grow on a substrate of threefold and fourfold symmetry respectively. On a substrate with twofold symmetry elongated island growth appears to be preferable[4,5].

For nanosturctural fabrication, usually a thin epilayer is grown on a substrate. This two-dimensional (2-D) layer is used to fabricate lower dimensional structures like wires (1-D) or dots (0-D) by lithographic techniques. However, structures smaller than the limits of conventional lithography techniques can only be obtained by *self-assembled* growth utilizing the principles of SK or VW growth. Due to their small size these islands can exhibit unusual quantum confinement effects which make them useful in electrical devices as single electron transistors and optical devices[6]. For appropriate growth conditions self-assembled epitaxial islands can be grown in reasonably well-controlled sizes[7,8]. A narrow size distribution is essential for most optoelectronic applications. The ability to grow such size-controlled islands is an attractive feature of self-assembly although the positions of individual islands cannot be determined precisely. Since precise placement of islands is not required for many devices, self-assembly is an attractive route for the fabrication of devices based on nanostructures.

Growth of Ge on Si is a classic case of Stranski-Kranstanov growth where uniform layer-by-layer growth up to ~3 monolayers [1 monolayer (ML) is equivalent to 6.78×10^{14} atoms/cm^2 on a Si(100) surface] is followed by island growth. Surface free energy of Ge is lower than that of Si. So Ge wets the Si surface – a condition for layer-by-layer growth. Both Ge and Si have diamond structure but different sizes of the unit cell (lattice constant: a^{Ge}=0.5657 nm, a^{Si} = 0.5431 nm). Due to 4.2% lattice mismatch between Ge and Si, a pseudomorphic Ge film grows with a strain. As the strain energy builds up, the competition between the surface energy and the strain energy eventually causes the film to undergo a 3-D island growth at coverages beyond 3 ML[9]. 3-D island formation leads to a partial relaxation of strain. During growth islands undergo a rich sequence of shape changes. On a Si(100) surface initially they are square or rectangular shaped. They are termed as hut clusters[10]. There are also other shapes like pyramid and dome[11]. All these islands can be dislocation free or coherent[9]. Large islands usually contain misfit dislocations, the presence of which reduces the strain energy. The islands, self-assembled via the SK growth mode, are in general of nanometric dimension.

2. Experimental

Ge was epitaxially grown on Si(100) and Si(110) surfaces by molecular beam epitaxy (MBE) in a compact custom-made MBE system and ***in-situ*** characterizations by reflection high energy electron diffraction (RHEED) and scanning tunneling microscopy (STM) were performed. MBE growth and characterizations by RHEED, STM and ***ex-situ*** TEM were performed in the following sequence. Si(100) samples were cleaned removing the native oxide by direct current heating at $\sim$ 1250°C. On this surface RHEED measurements were made to obtain the characteristic (2×1) reconstruction of the clean Si(100) surface. A thin (1.2 nm) Ge epilayer was grown by evaporating Ge from a Knudsen cell (Al_2O_3 crucible) at a deposition rate of 0.05 nm/min with the substrate kept at 550°C. Layer thickness was directly measured during deposition by the quartz microbalance. The thickness was calibrated by post-deposition measurement using Rutherford backscattering spectrometry (RBS). Base pressure in the MBE chamber was 1.0×10^{-10} mbar. During the growth the pressure increased to 8.5×10^{-10} mbar. During the deposition, RHEED measurements were also made. After the deposition had been completed, the samples were transfered to the STM chamber under ultrahigh vacuum condition. The STM chamber pressure was 2.5×10^{-10} mbar. Having completed the STM measurements, the samples were taken out of the vacuum chamber for TEM measurements. Plan view and cross-sectional TEM studies on identically prepared Ge/Si samples were carried out ex-situ. Identical methods were followed for growth on Si(110) surfaces.

3. Results and Discussions

RHEED pattern from a clean Si(100) surface, measured with a 15 keV electron beam, is shown in Fig.1(a). This shows a (2×1) surface reconstruction, which is the signature of a clean Si(100) surface. Fig.1(b) and Fig.1(c) represent the changes in the RHEED pattern during the MBE growth of Ge on the Si(100) surface. For a two-dimensional crystal the reciprocal lattice points are actually lattice rods. The Ewald's sphere for the 15 keV electron beam impinging on the surface at a grazing incidence (3°) cut the surface

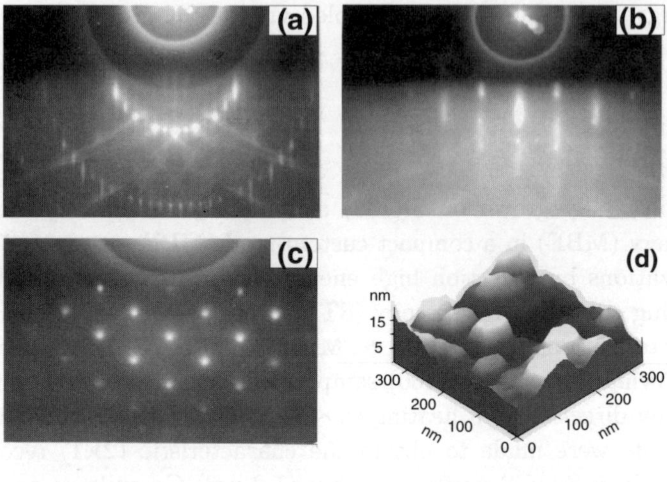

Figure 1. RHEED patterns from (a) Si(100)-(2×1) surface, (b) Ge/Si(100) during growth and (c) Ge/Si(100) after deposition of 1.2 nm Ge. (d) An STM image showing Ge islands for 1.2 nm Ge deposition.

reciprocal rods in elongated lines. This produces streak shaped diffraction spots as seen in Fig.1(a) for (2×1) reconstructed Si surface and in Fig.1(b) for a 2-D Ge layer. This is the signature of initial Ge growth as layer-by-layer. As growth continues, islands start growing beyond a deposition of 3–4 ML (0.4–0.5 nm) and 3-D island growth becomes dominant. Hence the streaked RHEED spots evolve into points as expected for 3-D growth. This is seen in the RHEED pattern in Fig.1(c) when 1.2 nm Ge has been deposited. The change in RHEED pattern from Fig.1(b) to 1(c) indicates a 2-D to 3-D transition as expected in the Stranski-Krastanov growth.

Having stopped the growth, the Ge/Si(100) sample has been transfered to the STM chamber for further studies. This transfer is done without disturbing the UHV condition of the system. In Fig.1(d) an STM image from this sample is shown. Nanometer scale Ge islands are now clearly seen in the real space. The islands are of square and rectangular shapes (although they have some rounded facets). Square island growth is expected from the fourfold symmetry of the Si(100) surface. However, rectangular shape is also possible through a shape transition[2]. These large islands apparently have flat top. Smaller islands can have 'hut' shapes[10]. With increasing coverage one also gets 'pyramid' and 'dome' shaped islands[11]. It is possible to get coherent islands upto a maximum size, beyond which islands grow with dislocations at the interface. However, whether the islands have dislocations or not can not be observed in the STM images. High resolution cross sectional transmission electron microscopy (XTEM) is more suitable for this purpose.

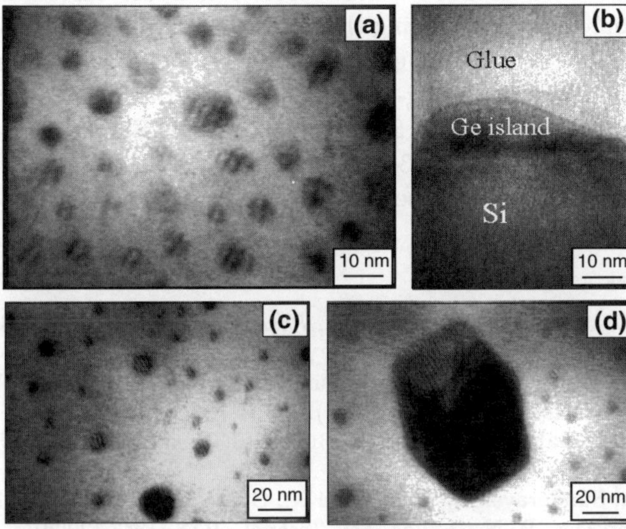

Figure 2. TEM micrographs: (a) plan view – Ge islands on Si(100) are seen. The image was taken from a region with a lot of small islands. Larger islands are also present. Cross sectional (XTEM) image of one such larger Ge island is shown in (b). Plan view image shows (c) Ge islands on a Si(110) surface, (d) a large Ge island on Si(110).

TEM micrographs for two samples are shown in Fig.2 [(a) and (b) for Ge/Si(100); (c) and (d) for Ge/Si(110)]. The plan view image in Fig. 2(a) shows Ge islands with some moire-fringes on the islands. A detailed study of the spacing, variation of spacing within an island and discontinuity of the moire-fringes can give evidence for the presence of misfit dislocations, nonuniform distribution of strain in an island and dislocation threading to the island surface[12,13,14]. Fig. 2(b) shows a cross-sectional image of a Ge island on Si and the Ge/Si interface. Images of many such islands have been used to study the Ge island height and shape variation as a function of substrate orientation and other growth parameters[15]. Fig.2(c) and 2(d) show Ge islands on a Si(110) substrate. On Si(110) surfaces growth of many very large islands along with smaller islands, was also observed. One such island is seen in Fig.2(d). We have also obtained lattice images in XTEM for studying the structure and defects across the Ge/Si interface. These will be reported elsewhere.

4. Conclusions

Recently, we have installed a cost-effective compact custom-made MBE system in our laboratory. Using this facility we have grown self-assembled epitaxial nanostructures of Ge on Si(100) and Si(110) substrates. We have also presented various during-growth and post-growth characterizations of these structures by RHEED, STM and TEM. RHEED patterns during Ge growth clearly showed 2-D→3-D transition, when following the completion of the wetting layer islands begin to grow.

References

1. Bauer, E. and van der Merwe, J. H., *Phys. Rev.*, **B33**, 3657 (1986).
2. Tersoff, J. and Tromp, R. M., *Phys. Rev. Lett.*, **70**, 2782 (1993).
3. Sekar, K., Kuri, G., Satyam, P. V, Sundaravel, B., Mahapatra, D. P. and Dev, B. N., *Phys. Rev.* **B51**, 14330 (1995).
4. Roeder, H., Hahn, E., Brune, H., Bucher, J. P. and Kern, K., *Nature*, **366**, 141 (1993).
5. Rout, B., Sundaravel, B., Das, Amal K., Ghose, S. K., Sekar, K., Mahapatra, D. P., and Dev, B. N., *J. Vac. Sci. Technol.*, **B18**, 1847 (2000).
6. Alivisatos, A. P., *Science*, **271**, 933 (1996).
7. Leonard, D., Krishnamurthy, M., Reaves, C. M., Denbaars, S. P. and Petroff, P. M., *Appl. Phys. Lett.*, **63**, 3203 (1993).
8. Drucker, J. and Chapparro, S., *Appl. Phys. Lett.*, **71**, 614 (1997).
9. Eaglesham, D. J. and Cerullo, B., *Phys. Rev. Lett.*, **64**, 1943 (1990).
10. Mo, Y. W., Savage, D. E., Swartzentruber, B. S. and Lagally, M. G., *Phys. Rev. Lett.*, **65**, 1020 (1990).
11. Medeiros-Ribeiro, G., Bratkovski, A. M., Kamins, T. I., Ohlberg, D. A. A. and Williams, R. S., *Science*, **279**, 353 (1998).
12. Hiroyama, Y. and Tamura, M., *J. Vac. Sci. Technol.* **A16**, 2956 (1998).
13. LeGoues, F. K., Copel, M. and Tromp, R., *Phys. Rev. Lett.*, **63**,1826 (1989).
14. Zou, Liao, X. Z., Cockayne, D. J. H., and Jiang, Z. M., *Appl. Phys. Lett.*, **81**,1996 (2002).
15. Satpati, B., Goswami, D. K., Satyam, P. V. and Dev, B. N. (unpublished).

RAMAN SPECTROSCOPIC STUDIES ON ELASTIC STRAIN AT GERMANIUM PARTICLES-SILICON MATRIX INTERFACE

ANUSHREE ROY AND SANGEETA SAHOO

Department of Physics and Meteorology,
Indian Institute of Technology, Kharagpur – 721302, INDIA
E-mail: anushree@phy.iitkgp.ernet.in

Germanium (Ge) clusters show luminescence at room temperature. The clusters are grown on Silicon (Si) substrate at room temperature (Ge-RT) and also at liquid nitrogen temperature (Ge-LNT) by cluster evaporation technique. Ge nanoparticles on glass substrates have been characterized by X-ray diffraction (XRD) measurements. The XRD pattern shows that Ge-LNT is not of diamond structure, as in bulk, but is tetragonal. The objective of the paper is to understand the effect of thermal annealing on both interfacial strain and interdiffusion of elemental Si at the interface, together with luminescence characteristics of the clusters. Raman measurement demonstrates the increase in strain with annealing in diffused disordered Si at the interface between Ge-LNT clusters and Si substrate. The decrease in PL intensity for Ge-RT with annealing has been attributed to reduction in surface oxide species, which is supported by Raman spectroscopic measurements.

1. Introduction

The clusters of group four semiconductor elements, silicon and germanium, grown on different matrices, are the two most important materials in microelectronics and optoelectronics industries. The presence of an indirect bandgap in these materials, render them useless for any optoelectronic device fabrication. In recent time a major research is going on in this field. The recent observation of visible photo- and electroluminescence at room temperature in Ge clusters on Si matrix, grown by cluster evaporation technique, has stimulated lot of excitement because of its potential optoelectronics applications [1]. The formation of these clusters may be understood as a complex interplay between cluster-matrix strain, surface energy, surface migration, and stress transport [1-3]. The luminescence properties of these clusters have been addressed in light of the annealing temperature during the cluster growth. Craciun et al. have reported that prolonged oxidation of Ge clusters on SiO_2 matrix results in large curvature at the interface due to high stress level induced by the volume expansion associated with the oxidation of the Si at the interface [4].

In the present report, we have discussed the effect of annealing in the Ge clusters grown on Si matrix at room temperature and also at liquid nitrogen temperature by cluster evaporation technique. We have characterized the clusters by X-ray diffractometry. As-prepared Ge nanostructures exhibit blue light emission. While photoluminescence (PL) spectra show the strain-relaxation in annealed Ge clusters, Raman measurements indicate corresponding increase in strain in diffused disordered Si layer at the interface between these clusters and Si substrate. Here we like to mention that because the broad 2TA (X_3) mode of Si is very close to the most intense LTO ($\Gamma_{25'}$) mode of Ge, it is difficult to study the effect of strain relaxation in the Ge clusters directly by Raman measurements. The objective of the paper is to understand the effect of thermal annealing on both interfacial strain and diffusion of Si at the interface together with luminescence characteristics of the Ge clusters.

2. Experiments

We have grown Ge clusters by cluster evaporation technique. In this technique the capped graphite crucible, loaded with small pieces of undoped Ge, is heated by indirect heating. The detail of the sample preparation condition has been discussed before [5]. The clusters of atoms are formed during adiabatic expansion in the nozzle of the crucible. The p-type (100) Si substrate is kept either at room temperature or at liquid nitrogen temperature. In this report, we represent the Ge clusters deposited at room temperature as Ge-RT and the same deposited at liquid nitrogen temperature as Ge-LNT.

X-ray diffraction patterns are obtained at room temperature by Philip's diffractometer, model PW17 with Cu K_α as the radiation source. For X-ray measurements samples are prepared on glass substrate. Photoluminescence spectra of the clusters are obtained at room temperature using Jonbin-Yvon fluorometer, model Fluoromax-3 with 313 nm as excitation wavelength. The Raman spectra are recorded at room temperature in 45^0 reflection geometry using 5145 Å line of Argon ion laser as excitation source and computer controlled SPEX Ramalog (model 14018) equipped with cooled photomultiplier tube and photon counting system as the detector. For thermal treatment in air, the samples are introduced into a temperature controlled furnace, maintained at a desired annealing temperature, and kept for 30 minutes. Then they are cooled at room temperature to record Raman and PL spectra. The samples are annealed at 4 different temperatures (T_a) ranging from 150 ^{0}C to 700 ^{0}C. The experiments are repeated on 6 samples to check the reproducibility of the results.

3. Results and Discussion

The XRD patterns of Ge-RT and Ge-LNT are shown in Fig. 1(a) and 1 (b), respectively. Both the patterns show broad glassy amorphous background. For Ge-LNT a sharp peak at $2\theta=33.68^\circ$, which can be associated with tetragonal phase from Ge clusters [1], rides on this broad background. For Ge-RT clusters this peak is very weak, shown in inset of Fig. 1 with 10 times magnification.

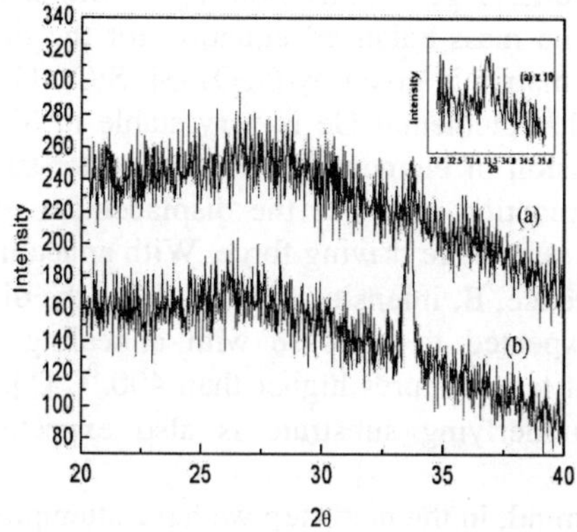

Fig. 1. XRD pattern for a) Ge-RT and b) Ge-LNT.

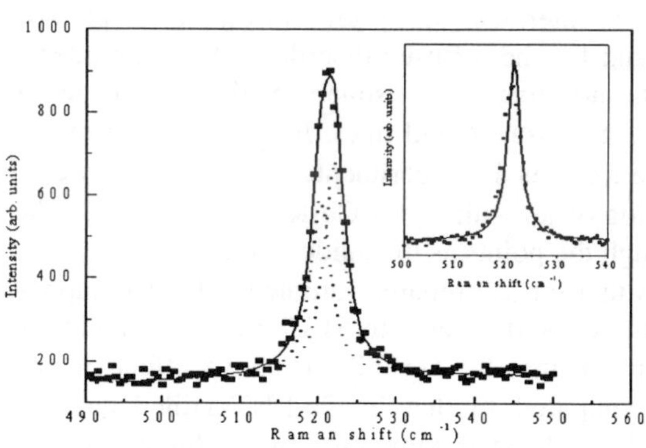

Fig. 2. Raman spectrum for 300^{0}C annealed Ge-RT are shown by dots.

We have studied the change in Raman spectra associated with Si, the matrix material on which Ge clusters are grown, as a function of annealing temperatures (150 ^{0}C, 300 ^{0}C, 500 ^{0}C, and 700^{0}C). To determine the FWHM of the spectrum from the bulk Si, under identical experimental conditions as for Ge-cluster samples, we have recorded Raman spectrum from p-type (100) Silicon wafer, which is the substrate material to grow the sample. We have tried to fit all the experimental Raman spectra of Si from our sample using a Lorentzian profile. In the inset of Fig. 2, we have shown the best fit obtained for Si from Ge-RT sample annealed at 300 ^{0}C. It is obvious from the inset that single Lorentzian alone cannot explain the Raman line shape from the Si in the sample. The same is true for the samples annealed at different T_a.

Both Ge-RT and Ge-LNT get oxidized during growth process and form GeO_x at the interface [1]. The mass-balanced equation for the displacement reaction between GeO_2 and Si matrix is given by GeO_2+Si=SiO_2+Ge. Thermodynamical consideration shows that elemental Ge is very stable in SiO_2 network and this reaction requires diffusion of elemental Si in the cluster surface [6-8]. If GeO_2 can obtain Si at the reaction interface the displacement reaction can proceed rapidly because of the very large driving force. With annealing in air the samples get oxidized further; hence, B, intensity of $I_B(\omega)$, from the diffused disordered Si at the interface is expected to increase with annealing. However, GeO_x is volatile, particularly at temperatures higher than 400 ^{0}C [1]. Hence, A, intensity of $I_A(\omega)$, from the underlying substrate is also expected to increase with annealing.

Having above in mind, in the next step we have attempted to fit the observed line shape of the same spectrum with $I(\omega) = I_A(\omega) + I_B(\omega)$, where $I_A(\omega)$ and $I_B(\omega)$ are the two Lorentzian line shapes arise from the Si substrate and disordered Si formed at the interface, respectively. Such strained peak has been observed at SiGe/Si interface [9,10]. ω_A and ω_B are the phonon frequencies for the Si substrate and for the diffused disordered Si at the interface, respectively. We have used the non-linear least square method to fit the data. In Fig. 2 we have plotted the fitted $I(\omega)$ (solid line). It can be seen from the Fig. 2 that the Raman line shape $I(\omega)$ fits the experimental spectrum very well. In Fig 3 we have shown the variation of ω_B with T_a for Ge-RT and Ge-LNT. The solid lines have been drawn through the points as the guide to the eye.

Fig. 3 shows that for as-prepared sample Ge-LNT ω_B appears at 521.3 cm^{-1}, with annealing this value decreases to 518.7 cm^{-1} monotonically (filled squares). We have also shown that for Ge-RT the above effect of annealing is not so pronounced as compared to that in Ge-LNT (filled circles). ω_B for Ge-RT remains essentially unchanged with annealing within experimental uncertainties. ω_A is unchanged for both Ge-RT and Ge-LNT as it arises from the bulk Si substrate.

The decrease of ω_B in Fig. 3 for Ge-LNT with increase in annealing temperature indicates increase in strain at the diffused disordered Si layer at the interface. It has been shown that as-prepared Ge-LNT contains more GeO_x than as prepared Ge-RT [1]. So from the above mass-balance displacement reaction one expects that the diffusion of elemental Si at the interface, hence the effect of $I_B(\omega)$ with annealing to be more in Ge-LNT than in Ge-RT because of higher concentration of GeO_x in the former, and the effect of strained diffused disordered layer shows up in Raman spectrum.

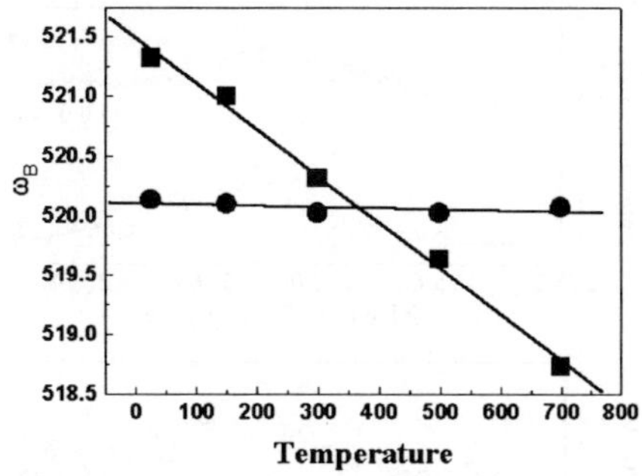

Fig. 3. Variation of ω_B (a) filled squares for Ge-LNT and (b) filled circles for Ge-RT room the bulk Si substrate.

We have also recorded corresponding PL spectra for as-prepared and annealed (300 ^{0}C, 500 ^{0}C, and 700^{0}C) Ge-RT and Ge-LNT. Fig. 4(a) and (b) show PL spectra of as-prepared and 500 ^{0}C annealed Ge-RT and Ge-LNT, respectively. The PL peak energies are estimated by fitting each spectrum by Gaussian functions. Three peaks appear at 2.61 eV, 2.62 eV, and 2.65 eV. As shown in Fig. 4 intensities of the PL peaks decrease with annealing.

The observed PL peak energies from the clusters in Fig. 4 are much higher than expected PL peak energy from crystalline Ge (0.67 eV). We attribute these PL peaks to surface oxidized Ge clusters, and surface species. Similar observation for the Ge clusters has been reported by Okamoto and Kanemitsu [11]. Choi et al. have explained the increase in photoluminescence intensity from Ge clusters, grown by molecular beam epitaxy technique, during annealing as due to the compressive strain on We suggest that the decrease in PL intensity in Ge-LNT with annealing, shown in Fig. 4(b), is not only due to the decrease in surface oxide species but also due to the strain-relaxation at the interface in these clusters, a likely result of the increased strain in diffused clusters by the matrix [3]. In Fig. 4(a) the decrease in PL intensity in Ge-RT can be attributed to reduction in surface oxide species with annealing.

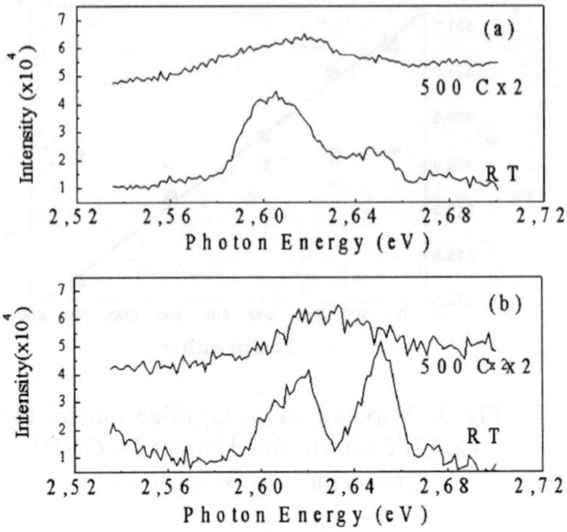

Fig. 4. Photoluminescence spectra of
a) as-prepared and 500^0C annealed Ge-RT
b) same for Ge-LNT.

4. Conclusion

In this paper, we have discussed effect of annealing on strain-relaxation in Ge clusters grown on Si substrate at liquid nitrogen temperature and also at room temperature by cluster evaporation technique. While PL measurements help in understanding the nature of strain relaxation in Ge-LNT, Raman scattering probes the corresponding increase of strain in the diffused disordered Si at the interface. Effect is not pronounced for the clusters grown at room temperature.

References

1. K.P.Jain (Ed), *Physics of Semiconductor Nanostructures*, Narosa Publishing Home. p 154; S. Sato, S. Nozaki and H. Morisaki, *Appl. Phys. Lett.*, "Tetragonal Ge films deposited by the cluster-beam evaporation technique", **66** (1995) pp 3176-3178.
2. H.Chen, W.Q. Cheng, X.G. Xie, Q. Huang, and J.M. Zhou, "Optical transition in SiGe self-organized dots", *Appl. Phys. Lett.*, **70** (1997) pp 446-448.

3. W.K.Choi, V. Ng, S.P. Ng, H.H. Thio, Z.X. Shen and W.S. Li, " Ranan investigation of Ge nanocrystals in amorphous Si oxide films synthesized by rapid thermal annealing", *J. Appl. Phys.*, **86** (1999) pp 1398-1403.
4. V. Craciun, C. Boulmer-Leborgne, E.J. Nicholls, I.W. Boyd, "Light emission from Ge nanoparticles formed by ultraviolet oxidation of silicon-germenium", *Appl. Phys. Lett.*, **69** (1996) pp 1506-1508.
5. Elastic strain at the interface of Ge clusters and Si interface –A Raman spectroscopic measurement, J. Phys. Chem Solids (in press).
6. Y. Maeda, "Visible photoluminescence from nanocrystalline Ge embedded in a glassy SiO_2 matrix: Evidence in support of the quantum confinement mechanism," *Phys. Rev. B*, **51** (1995) pp 1658-1670.
7. D. C. Paine, c. Caragianis, and A.F. Schwartzman "Oxidation of SiGe alloys atmospheric and elevated pressure", *J. Appl. Phys.*, **70** (1991) pp 5076-5084.
8. N.R. Zangenberg, J.Lundsgaard Hansen, J. Fage-Pedersen, and A. Nylandsted Larsen, "Ge self-diffusion in epitaxial Si_{1-x} Ge_x layers", *Phys. Rev. Lett.*, **87** (2001) pp 125901-1-125901-4.
9. B. Dietrich, E. Bugiel, H.J. Osten, and P. Zaumseil, "Raman investigation of elastic strain relief in Si_{1-x} Ge_x layers on patterned Si substrate", *J. Appl. Phys.* **74** (1993) pp7223-7227.
10. S.J. Koester, K. Rim, J.O. Chu, P.M. Moony and J.A. Ott, M.A. Hargrove, "Effect of thermal processing on strain relaxation and interdiffusion in Si/SiGe heterostructures studied using Raman spectroscopy", *Appl. Phys. Lett*, **79** (2001) pp 2148-2150.
11. S. Okamoto and Y. Kanemitsu, "Photoluminescence properties of surface-oxidized Ge nanocrystals: surface localized of excitation", *Phys. Rev.* **B54** (1996) pp 16 421-16 424.

LAYERED SYNTHETIC MICROSTRUCTURES

LAYERED SYNTHETIC MICROSTRUCTURES: IMPORTANCE OF A COMBINED X-RAY STANDING WAVE AND X-RAY REFLECTIVITY ANALYSIS

B. N. DEV

Institute of Physics, Sachivalaya Marg, Bhubaneswar-751005, India

Layered synthetic microstructures consisting of thin layers of alternating elements or compounds have unique structural, electronic and magnetic properties with a wide range of applications. For such multilayer systems it is important to correlate the measured properties with structure and composition so that preparation techniques can be optimized to yield high performance materials. For magnetic multilayers where alternating layers are magnetic and nonmagnetic materials, a small amount (even a few percent) of magnetic impurity (either from the magnetic layers or external) in the nonmagnetic layer can change the magnetic coupling and magnetoresistance significantly. Determination of such small concentrations by X-ray reflectrometry is often very difficult. Layer composition is more accurately determined by X-ray standing wave (XSW) analysis, while interface roughness is more accurately determined by the X-ray reflectivity (XRR) technique. It is shown that for periodic multilayers, a combined XSW and XRR analysis provides more accurate compositional and microstructural information overcoming the deficiencies of the individual techniques.

1. Introduction

Layered synthetic microstructures (LSM's), consisting of thin layers of alternating elements or compounds (A/B/A/B....), have unique structural[1], magnetic[2] and electronic[3] properties with a wide range of applications. However, their properties depend strongly on the structural details, such as the thickness ratio of A and B layers, roughness of the A/B and B/A interfaces, impurities in the respective layers – either due to interdiffusion or from external sources and the total thickness of the A/B/A/B.... stack. These structural parameters govern the performance of multilayers in various applications, such as X-ray astronomy, microscopy, spectroscopy, monochromators etc. For magnetic multilayers, where A is a magnetic material and B is nonmagnetic or vice versa, small concentrations of magnetic impurities in the nonmagnetic layers can drastically change magnetic coupling and megnetoresistance[4]. This highlights the role of layer composition.

Spin reorientation transition can be caused by changing structural parameters of a magnetic multilayer[5]. When this change of parameters are made in a spatially modulated fashion, such modifications can lead to fabrication of patterned ultrahigh density recording media[5,6,7]. It is extremely important to analyze the structural and compositional details of LSM's in order to understand their properties and improve their performance.

Most of these LSM's have layer thicknesses in the nanometer length scale. There are not many experimental techniques which can probe structural and compositional details at this length scale. From X-ray reflectivity (XRR) experiments it is possible to determine layer thickness, interface roughness and periodicity of a periodic multilayer with high precision. However, small impurity concentrations cannot be reliably determined by XRR. Secondary ion mass spectrometry (SIMS) has very high sensitivity and, in principle, is capable of determining very small impurity cocentrations. However, SIMS has its limitation on depth resolution ($\sim$ 50 Å). X-ray standing wave (XSW) technique is capable of determining a small impurity concentration with high depth resolution. However, XSW cannot reliably determine the roughness parameters. A combined analysis of periodic multilayers by XRR and XSW can, on the other hand, provide a more detailed structural and compositional information[8].

The role of small impurity concentrations in multilayers is evident from the example of a Co/Cu multilayer where the nonmagnetic Cu layers contained Ni impurity. As the Ni concentration becomes smaller, magnetic coupling and magnetoresistance show a more interesting behaviour. In the work of Parkin *et al.*[4] on Co(10 Å)/$Cu_{1-x}Ni_x$ (4 – 50 Å) multilayers, for the smallest concentration of 4% Ni in Cu, the system shows the largest magnetoresistance. So the natural question is, would the effect be stronger for even smaller concentrations? This also necessitates the determination of this concentration accurately. The quoted concentration in ref.4 appears to be the nominal concentration expected from the preparation. This does not guarantee that after sample preparation the concentration of Ni in the Cu layer remains as expected. Diffusion across the interface can change the concentration. This is spectacularly obvious for Au/Cu multilayers, where Cu and Au were found to be completely intermixed in about a month after the fabrication of the multilayer[9]. While this is an extreme case of interdiffusion, in other cases the diffusion may be slow and the layer composition may change slowly. Hence the detection of a small concentration ($\sim$ 1%) of impurities by a suitable technique is desirable. XSW is a suitable technique for this purpose. Combined XSW and XRR can be used to determine structure and composition of the LSM's. This has been shown in details in ref.8, where a Pt/C multilayer sample has been analyzed and the composition of the C layers was found to be $Pt_{0.05}C_{0.95}$. From the fabrication of the multilayer the C-layers were intended to have no Pt. Other Pt/C samples with different Pt-to-C thickness ratios, showed somewhat different concentrations[10]. These $Pt_xC_{(1-x)}$ compositions of the C-layers were not intended during the Pt/C multilayer fabrication. Apparently, diffusion at room temperature has changed the composition of individual layers, which would otherwise be pure Pt and C layers.

In order to substantiate the claim that the combined XSW and XRR technique is capable of determining the impurity concentration with $\sim 1\%$ precision, it is necessary to introduce a small known change of concentration and then determine the change experimentally. This is difficult to ensure during the growth of the multilayer as we have observed earlier[8,9,10] that there can be an unintended change of a few percent even in the most stable multilayers. We have taken an alternative route using an athermal process to displace some Pt atoms from the Pt layers into the C layers and vice versa. This has produced a known concentration change which has been measured by the combined XSW and XRR technique. Before we take up examples we discuss the deficiencies of XRR and XSW techniques when they are applied individually. Then the relevant points are clarified using our previously published results.

2. Deficiency and strength of the XRR technique

The analysis of the X-ray reflectivity data is based on the electron densities of the individual layers[11]. It does not directly reveal the identity of the elements present. The same electron density can arise from many possibilities. Multilayers, usually fabricated by sputtering in an inert gas atmosphere, say Ar, are known to incorporate a small amount of Ar within the layers[12]. This has an effect of changing the electron density of the layers, besides any change due to interdiffusion. If incorporation of impurity in a layer causes volume change of the layer, it may also introduce an electron density change in an unpredictable manner. Thus from the estimate of the electron density, it is difficult to obtain correct impurity concentration. Even a worse situation arises when the change in electron density due to impurity incorporation is so small that it does not produce a detectable difference in reflectivity. Let us take the case of Ni impurity in Cu as the case in ref.4. Ni and Cu has nearly equal electron density and a small amount of Ni in Cu is not expected to change the electron density of a Cu layer significantly. For a Pt/C multilayer, an example showing the effect of a reasonably large change of electron density on reflectivity will be presented below.

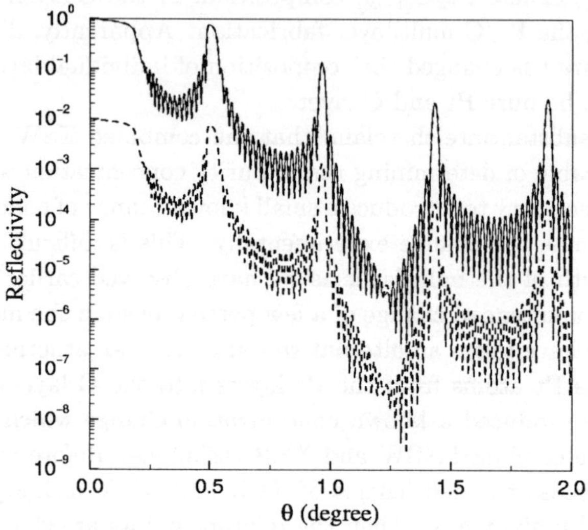

Figure 1. Theoretical plots of reflectivity for a 20–period Pt/C multilayer system for different electron densities ρ_C of the C layers. $\rho_c = 0.698e/\AA^3$ (—), and $\rho_c = 0.803e/\AA^3$ (15% higher compared to the actual density) (- - -). Curves are vertically shifted by two orders. However, they are also shown in an overlapping mode to demonstrate that they are practically indistinguishable. (From ref.8)

In the case of a Pt/C multilayer, some dissolved Pt in the C layers would increase the electron density of the C layers. Two reflectivity plots are shown in Fig.1. In both cases Pt electron density is the same ($\rho_{Pt} = 5.05e/\AA^3$). However, the carbon layer electron density ($\rho_C = 0.698e/\AA^3$) is 15% higher ($\rho_C = 0.803e/\AA^3$) in the second case. The reflectivity plots are hardly distinguishable. Thus determination of a small Pt concentration in the C layers in such a case would be nearly impossible from an XRR analysis.

The strong point of XRR remains the determination of surface and interface roughness. Theoretical plots in Fig.2 show the strong sensitivity of reflectivity to roughness for a Pt/C multilayer.

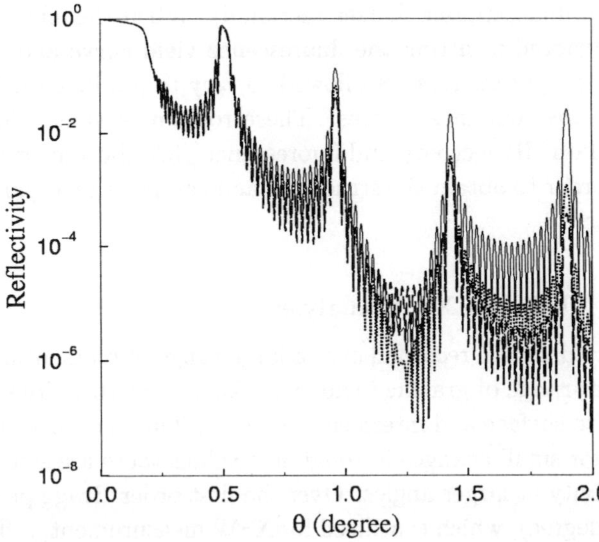

Figure 2. Reflectivity from a 20-period Pt/C multilayer system with periodicity d (43 Å)= d_1 (17 Å)+d_2 (26 Å), and with surface and interface roughnesses (Å) σ_0, σ_1, σ_2: 0, 0, 0 (—); 3, 3, 3 (.....) and 3, 5, 3 (- - -) (From ref.8).

3. Deficiency and strength of the XSW technique

In the XSW technique, reflectivity is measured across a Bragg peak (usually first order) and simultaneously an inelastic signal, such as fluorescence, from one or more types of atoms is detected. This carries the identity of the atom that emit the inelastic signal due to incident X-rays. Neither the reflectivity over the small angular range of the Bragg peak nor the yield of the inelastic signal is strongly sensitive to surface and interface roughness.

On the other hand, depth distribution of any type of atoms over the multilayer period and hence the impurity concentration can be determined more precisely by the XSW technique.

Kawamura and Takenaka[13] tried to determine the interface roughness from XSW experiments on Ni/C multilayers by fitting Ni K_α fluorescence yield. However, for the analysis of reflectivity they assumed the interface roughness to be zero. This is inconsistent and the extracted roughness values from the XSW experiments are not reliable. Ghose and Dev[8] have shown that the fluorescence yield in the XSW experiment can be fitted with interface roughness as parameters. However, these roughness parameters extracted from the fluorescence fit, when used for fitting the reflectivity data, produce an unacceptable fit (see section 4). When the effect of layer composition is ignored in fitting the fluorescence yield curve and only the interface roughness parameters are allowed to vary to produce the best fit, it overestimates the roughness values. These roughness values do not fit the reflectivity data. Reflectivity and fluorescence yield data must be fitted consistently in order to obtain the structural and compositional parameters of the multilayers[8].

4. Combined XRR and XSW analysis

X-ray reflectivity data are recorded over a large range of momentum transfer (large angular range of grazing incidence). Data over this large range is highly sensitive to surface and interface roughness. This is clear from Fig.2. We notice that for small changes in roughness values there are strong deviations in reflectivity at larger angles. Over the first order Bragg peak (here around $\theta = 0.5$ degree), which is utilized for XSW measurement, reflectivity is less sensitive to roughness.

In order to obtain correct structural and compositional parameters of a periodic multilayer system, the prescription is to fit the reflectivity data to extract roughness and other parameters[8]. For fitting the fluorescence yield data all the parameters obtained from the reflectivity fit are to be kept fixed and the coherent fraction[a] (f_c) is to be allowed to vary allowing the presence of a concentration of fluorescing atoms in the other layer. (For example, to fit Pt fluorescence data allow the presence of some Pt in the C layers). Then compute the average composition from the coherent fraction obtained from the best fit. The detailed procedure was explained in ref.8.

To illustrate this let us take a given case shown in Fig.3, where reflectivity fit (Theory-1) has given the following roughness values: σ_0(surface) = 3 Å, σ_1(Pt-on-C interface) = 4.5 Å and σ_2(C-on-Pt interface) = 2.9 Å. When the Pt fluorescence yield (Fig.4) was fitted allowing σ_0, σ_1 and σ_2 to vary, the best fit (with $f_c = 1$) yielded $\sigma_0 = 3$ Å, $\sigma_1 = 8.9$ Å and σ_2 = 4.2 Å. When this second set of roughnesses is used to fit the reflectivity data in Fig.3, the fit (Theory-2) becomes unacceptable. The fit over the first-order Bragg peak is not too bad, however, the deviations over higher order peaks are strong. It is clear that the roughness values extracted from

[a]For an A/B/A/B....multilayer, when fitting the fluorescence yield from A, the coherent fraction, f_c, indicates the fraction of material A remaining in the A layers including the broadened interfaces. The fraction $(1 - f_c)$ of material A is assumed to be in uniform depth ditribution in the B layers.

the fluorescence data fit are not consistent with those from the reflectivity data fit.

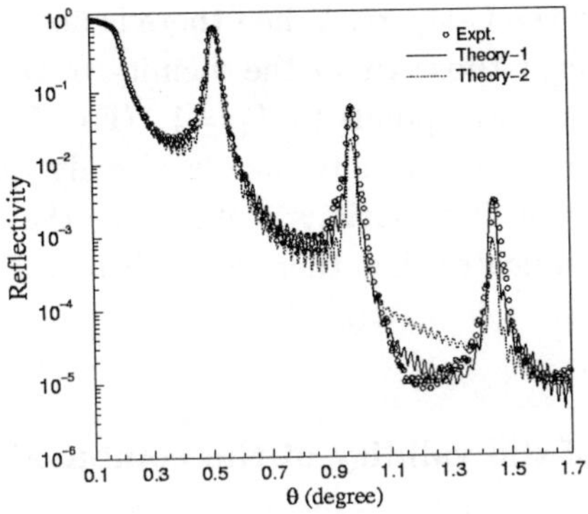

Figure 3. Experimental reflectivity data (o o o) and fitted theoretical reflectivity curve (—) for a Pt/C multilayer on a glass substrate with 20 bilayers. Parameters obtained from the fit: bilayer thickness d = 42.9 Å, Pt layer thickness d_1 = 16.8 Å, C layer thickness d_2 = 26.1 Å, surface roughness σ_0 = 3 Å, Pt-on-C interface roughness σ_1 = 4.5 Å, and C-on-Pt interface roughness σ_2 = 2.9 Å. The theoretical reflectivity curve (....) for σ_0 = 3 Å, σ_1 = 8.9 Å, and σ_2 = 4.2 Å; all other parameters are unchanged.

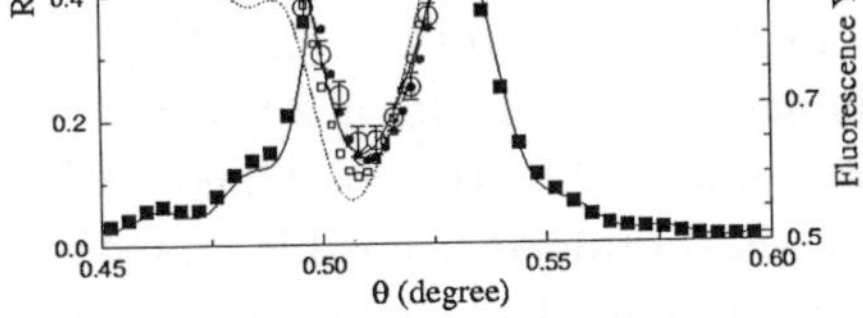

Figure 4. Experimental Pt L_α fluorescence yield (o o o) and reflectivity (solid squares) vs angle of incidence θ over the first order Bragg reflection and the theoretical curves. (....): σ_0 = 3 Å, σ_1 = 4.5 Å, σ_2 = 2.9 Å, and f_c = 1.0 (*i. e.*, no additional Pt in the C layers other than those near the interfaces). (—): σ_0 = 3 Å, σ_1 = 4.5 Å, σ_2 = 2.9 Å and f_c = 0.87 ($Pt_{0.05}C_{0.95}$). (- - -) σ_0 = 3 Å, σ_1 = 8.9 Å, σ_2 = 4.2 Å and f_c = 1.0. Also shown (for σ_0 = 3 Å, σ_1 = 4.5 Å and σ_2 = 2.9 Å) are the curves (open squares) for $Pt_{0.03}C_{0.97}$ (f_c = 0.935) and (filled circles) $Pt_{0.07}C_{0.93}$ (f_c = 0.844), which deviate considerably from the best fit for f_c = 0.87. (Both figures 3 and 4 are from ref.8).

On the other hand, when the roughness values extracted from the reflectivity data fit are kept fixed, the best fit of the fluorescence data (Fig.4) provided a coherent fraction of 0.87 which correspond to an average composition of $Pt_{0.05}C_{0.95}$ for the C-layers. When there is no additional dissolved Pt in the C layers, except those Pt in the vicinity of the Pt/C and C/Pt interfaces, this situation corresponds to $f_c = 1$. (For $f_c = 1$, contribution from Pt in the C-layers near the interface is already taken into account from error function depth distributions[8]; otherwise the C-layers are pure C). A value of $f_c < 1$ indicates that there is additional Pt in the C-layers, *i.e.*, $x \neq 0$ in $Pt_xC_{(1-x)}$.

5. Demonstration of the validity of the combined XSW and XRR approach

From Fig.4 it is seen that the precision in the determination of concentration is better than 2%. In order to demonstrate the validity of the combined XRR and XSW method and the precision on the determination of concentration, a controlled change in concentration in a Pt/C multilayer can be produced by ion irradiation and then determined by the combined XSW and XRR method. Earlier we investigated the effect of ion irradiation on Pt/C multilayers by the combined XRR and XSW analysis[14]. Results in ref.14, when considered in reversed logic, can serve as a demonstration of the precision on the determination of concentration. This is what is presented below using the results of ref.14.

Ion-atom collision in the Pt/C multilayer produces atomic displacements within the sample which displaces some Pt atoms into the C layers and vice versa. This changes the composion from $Pt_xC_{(1-x)}$ to $Pt_{(x+\delta x)}C_{(1-x-\delta x)}$ in the C layers and from $Pt_{(1-y)}C_y$ to $Pt_{(1-y-\delta y)}C_{(y+\delta y)}$ in the Pt layers. By ion irradiation a change of $\delta x \approx 0.026$ or 2.6% has been produced and the irradiated sample was analyzed by the combined XRR and XSW technique. A value of $\delta x = 0.02$ was obtained from this analysis.

In order to produce a controlled change in concentration by ion irradiation, Monte Carlo simulation was performed to obtain quantitative values of ion induced atomic displacements in the Pt/C multilayers. This is illustrated below.

Ion beam induced atomic displacements in matter is a fairly well understood subject in the ion energy regime considered here. For the simulation 2 MeV Au ions and three layer pairs of Pt(15 Å)/C(20 Å) on a C substrate were taken. Monte Carlo simulation code SRIM2000[15] was used to

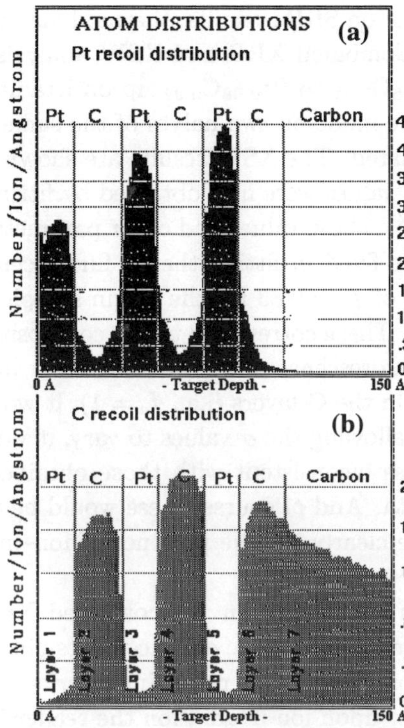

Figure 5. Distributions of displaced (a) Pt and (b) C atoms due to 2 MeV Au ion irradiation of a Pt/C multilayer, as obtained from a Monte Carlo simulation. Sharp Pt/C layer boundaries before ion irradiation are marked by vertical lines. Incorporation of displaced Pt in C layers and vice versa leads to Pt/C interface broadening and change of layer composition. (From ref.14)

calculate the number of displaced Pt and C atoms in each layer.

Results obtained for 2000 incident 2 MeV Au ions are shown in Fig.5. We notice several features from Fig.5. A large number of Pt atoms are displaced within the Pt layers. Some Pt are displaced across the Pt/C and C/Pt interfaces. This not only broadens the Pt distribution at the interfaces but also changes the average composition of the C layers. Fig.5 shows the same trend for C displacements. At the middle of the C-layers we notice on the average there are ~ 0.3 Pt atoms per incident ion per $\AA$ depth. When a fluence of 1×10^{14} ions/cm^2 is used in irradiation, 1×10^{14} incident ions within a 20 $\AA$ thick C-layer produce $\sim 6 \times 10^{14}$ displaced Pt atoms or $\sim 2.6\%$ Pt in the C layers. (Although actual concentration may be somewhat different because of the error on SRIM results, thermodynamic parameters like heat of mixing etc.). Combined XRR and XSW analysis shows a composition change from $Pt_{0.06}C_{0.94}$ to $Pt_{0.08}C_{0.92}$ up on irradiation. XRR and XSW measurements were made on two halves of the same sample, on which one half was ion irradiated. The XSW results are shown in Fig.6. In this analysis σ_0, σ_1 and σ_2 values were first obtained by fitting the reflectivity data (not shown here)[14]. These values and other parameters optimized from the reflectivity fit were fixed in fitting the Pt fluorescence data to obtain the coherent fractions of $f_c = 0.83$ for the virgin sample and $f_c = 0.80$ for the irradiated sample. These correspond to the compositions mentioned above. In fitting the Pt fluorescence data, if on the other hand, one had assumed no presence of Pt in the C-layers (i.e., $f_c = 1$), it would have been possible to fit the data by allowing the σ values to vary, it would have yielded large values of roughness inconsistent with those obtained from the best fit of the reflectivity data. And of course, these would be wrong results as, from the simulation, we clearly see the presence of ion–induced displaced Pt in the C layers.

All the results obtained from the combined XRR and XSW analysis can be explained from the atomic displacements obtained in the simulation shown in Fig.5. For other details regarding interface broadening, multilayer period dilation etc. upon ion irradiation the reader is referred to ref.14.

We have used the ion irradiation route to produce a change in composition for two reasons. First, the ion–solid interaction in this energy regime is fairly well understood so that an intended change in composition can be achieved fairly accurately. Secondly, there is a technological aspect of ion beam modification. Ion beam patterning of magnetic thin films through suitable stencil masks has been suggested recently as a prospective path towards patterned, ultrahigh-density magnetic recording media [6,7]. Magnetic multilayers are ideally suited for this application, since bombardment with a suitable fluence of energetic ions may induce interface mixing (as seen in Fig.5), thereby triggering a spin-reorientation transition[5]. Such ion irradiation induced transition has been observed for Co/Pt multilayers, which were found to undergo a spin-reorientation transition from easy axis out-of-plane to easy axis in-plane[5]. As we notice from Fig.5, there is not only interface mixing, but also a change in composition. In ref.4, we notice that the magnetic behaviour strongly depends on the composition of the nonmagnetic layers. In ion bombardment, both interface mixing and change of composition occur. This raises the question whether interface mixing or the average layer composition change or both are responsible for

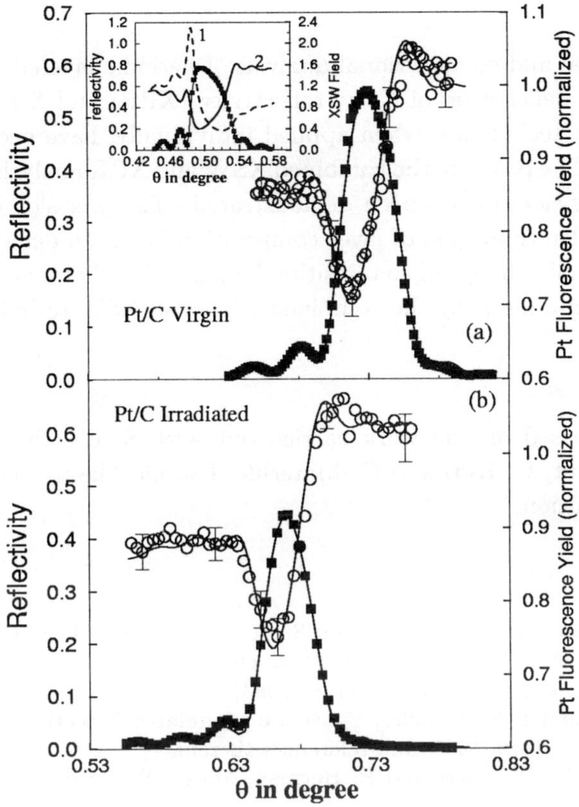

Figure 6. Experimental Pt L_α fluorescence yield (open circle) and reflectivity (solid squares) vs angle of incidence θ over the first order Bragg reflection and the theoretical curves (solid curve) for (a) virgin and (b) irradiated periodic Pt/C multilayer samples. Fluorescence data have been normalized at $\theta = 2.2°$. The position of the Bragg peak for the case of irradiated sample has shifted to a smaller angle because of the increase of the multilayer period. The inset in (a) shows the theoretical XSW field variation with θ over the C layers (1) and over the Pt layers (2), for the first order Bragg peak. Pt present in the C–layers produce Pt fluorescence following curve–1, while those in Pt layers follow curve–2. A change in Pt concentration in C–layers produce a change in the shape of the effective Pt fluorescence yield curve. The shape analysis via the parameter f_c gives the Pt concentration. (From ref.14).

the observed spin-reorientation transition. Further investigations will be necessary to answer this question. The combined XSW and XRR method would be ideal for exploring the microstructure and composition in the quest for understanding the structure–composition–property relationship in such multilayer systems.

6. Conclusions

For the structural and compositional analysis of periodic layered synthetic microstructures, deficiencies of X-ray reflectivity (XRR) and X-ray standing wave (XSW) techniques, when applied individually, have been shown with examples. The power of the combined XSW and XRR analysis in overcoming those deficiencies has been demonstrated. The precision of better than 2% in the determination of layer composition has been demonstrated by producing a 2.6% change in composition by ion-induced atomic displacements and determining it by the combined XRR and XSW technique.

Acknowledgement

This article is based on the work carried out with S. K. Ghose, D. K. Goswami, B. Rout, G. Kuri and G. Materlik. I would like to thank them for their contribution.

References

1. D. B. McWhan, Synthetic Modulated Structure, edited by L. L. Chang and B. C. Giesser, Academic, New York, 1985, chap. 2, p.43.
2. M. B. Stearns, J. Appl. Phys.**55**, 1729 (1984).
3. C. M. Falco and I. K. Schuller, Synthetic Modulated Structure Materials, Academic, New York, 1985, and references therein.
4. S. S. Parkin, C. Chappert and F. Herman, Mater. Res. Soc. Symp. Proc., **313**, 179 (1993).
5. D. Weller *et al.*, J. Appl. Phys., **87**, 5768 (2000).
6. C. Chappert *et al.*, Science, **280**, 1919 (1998); J. Ferre *et al.*, J. Magn. Magn. Mater. **198–199**, 191 (1999); T. Devolder *et al.*, Appl. Phys. Lett., **74**, 3383 (1999).
7. B. Terris *et al.*, Appl. Phys. Lett., **75**, 403 (1999).
8. S. K. Ghose and B. N. Dev, Phys. Rev. **B63**, 245409 (2001).
9. J. DuMond and J. P. Youtz, J. Appl. Phys., **11**, 357 (1940).
10. S. K. Ghose *et al.* (unpublished).
11. L. G. Parratt, Phy. Rev. **95**, 359 (1954).
12. T. Matsusita, A. Iida, T. Ishikawa, T. Nakagiri and K. Sakai, Nucl. Instrum. Methods Phys. Res., **A246**, 751 (1986).
13. T. Kawamura and H. Takenaka, J. Appl. Phys., **75**, 3808 (1994).
14. S. K. Ghose, D. K. Goswami, B. Rout, B. N. Dev, G. Kuri and G. Materlik, Appl. Phys. Lett., **79**, 467 (2001).
15. SRIM2000 – a version of the TRIM program: J. F. Ziegler, J. P. Biersack and U. Littmark, The Stopping and Range of Ions in Matter (Pergamon Press, New York, 1985).

DEVELOPMENT OF MULTILAYERS FOR HARD X-RAY OPTICS

YUZURU TAWARA, KOUJUN YAMASHITA, YASUSHI OGASAKA, KEISUKE TAMURA, KAZUTOSHI HAGA, TAKASHI OKAJIMA, AKIHIRO FURUZAWA, SEIMA KATO, HIROYUKI SATAKE, KENTARO NOMOTO, NORIYUKI HAMADA, SATOSHI TAKEUCHI, TSUTOMU MORISHITA

Dept. of Physics, Nagoya University
Furo-cho, Chikusa, Nagoya, 464-8602,
JAPAN
E-mail: tawara@u.phys.nagoya-u.ac.jp

We have been studying multilayers in the view point of developing X-ray optical instrument mainly for X-ray astronomy. Recently we successfully made hard X-ray telescope using platinum carbon multilayer supermirror with effective energy band up to several tens of keV for the first balloon experiment. In this telescope, we use a few hundred nested thin foil mirrors with radii of 60 - 200 mm, which correspond to grazing incidence angles of 0.11 - 0.36 degree for 8 m focal length. On the surface of each mirror, we deposited Pt/C depth graded multilayes, so-called supermirror by DC magnetron sputtering system. Multilayer parameters, such as d-spacing, ratio of Pt thickness to d-spacing, total number of layer pairs were searched in order to maximize integrated reflectivity in the given energy band. Evaluation of the X-ray reflection performance was done by X-ray beam line in our laboratory and synchrotron facility SPRING8. From these measurements, we found that the basic performance was as expected and especially the averaged interfacial roughness was 0.4 nm. In this paper, we will discuss general design of multilayer supermirror for hard X-rays, fabrication and evaluation of such multilayers and future applications.

1. Introduction

According to the development of thin film fabrication technology, it has become possible to make very thin and fine multilayers. For optics application, such multilayers are treated as artificial Bragg crystal with d-spacing designed by ourselves. Especially for hard X-rays of E > 10keV, there is a great progress in developing X-ray telescope using multilayer supermirror. From the viewpoint of the physics of surface and interfaces, there are several interesting problems in this development, such as minimum d-spacing, minimum interfacial roughness for a given combination of multilayer.

In this paper we will describe basics of multilayers (Section 2), hard X-ray telescope as an important application of such multilayers to the X-ray optics (Section 3), design of multilayer supermirror (Section 4), recent result of X-ray measurement of multilayer supermirror (Section 5) and summary and future prospect (Section 6).

2. Multilayer and multilayer supermirror

Multilayer, the layered synthetic microstructure, has multiple layer pairs of light and heavy element. Its periodic structure works as one-dimensional artificial Bragg crystal, which can reflect X-rays in the Bragg condition: $2d\sin\theta = n\lambda$, where d the bilayer thickness or so-called d-spacing, θ the grazing incidence

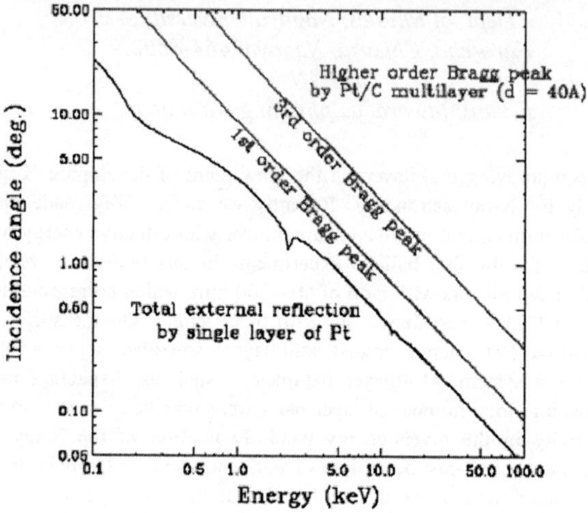

Figure 1. Effective region of energy and incidence angle for X-ray reflection

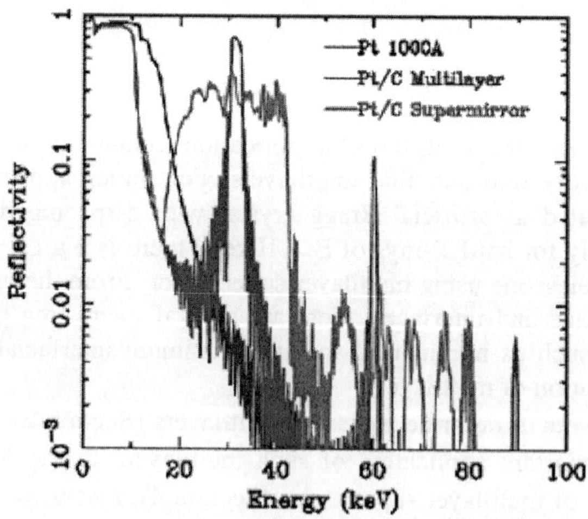

Figure 2. X-ray reflectivity of monolayer, multilayer and supermirror

angle, n the order of Bragg reflection and λ the wavelength of incident photon.

Since total external reflection can be used only in the parameter region shown in figure 1, it is the only way to use Bragg reflection to make X-ray reflector for X-rays above the critical energy or above the critical angle of incidence.

For example, if we use multilayer with d-spacing of 4 nm and incident angle of 0.5 deg, 20 keV X-ray can be reflected by Bragg reflection, which is about factor of 2 higher than critical energy. Although the high reflectivity at the Bragg peak can provide large effective area of X-ray telescope, it is also very important to have wide energy band, from the view point of astronomy to get maximum information of X-ray source in the universe.

In order to get such property for reflector, we use depth graded d-spacing multilayer, so-called supermirror as shown in figure 2. Basic idea of supermirror originated from neutron optics, where attenuation can be neglected. If we stack multilayer blocks having different d-spacing, we can expect Bragg peak x-ray reflectivity corresponding to each d-spacing. The actual image of such supermirror made by ourselves taken by transmission electron microscope shown in Fig. 3. In this case d-spacing and bilayer number range from 3.4 – 5.6 nm and 3 –15, respectively.

On the material selection of multilayer for X-ray optics, it depends on the kind of optical system and X-ray energy. In this paper we concentrate on Pt/C multilayer, since we selected them as a mirror material for hard X-ray telescope. One of the reasons why we select this combination is that platinum has relatively high electron density and thus high reflectivity can be achieved with small number of total layer pairs. This is important in the view point of mass production, since we should make thousands of mirrors with typical size of 10-20 cm square for one full telescope and it takes several hours on the average to deposit multilayer for each mirror. At the same time it may not be the best selection since platinum is not cost-effective for covering large area of mirror surface of several tens m^2 for one telescope.

Another material of carbon is chosen as a light element pair, since Pt/C multilayer can be made with relatively small interfacial roughness and small effect of interdiffusion as shown in figure 3.

Figure 3. TEM image of Pt/C multilayer

3. Application of multilayers to a hard X-ray telescope

One of the novel applications of multilayer is the hard X-ray telescope.

Figure 4 shows the telescope made by our group and NASA/Goddard Space Flight Center group prepared for the balloon experiment, "InFOCμs". This is the very first hard X-ray telescope using multilayer supermirror [1, 2, 3, 4, 5]

Diameter of this telescope is 40cm and there are 255 different radii nested thin foil mirrors as shown in Fig. 4 and Fig. 5. This telescope covers 20-40keV energy band and has effective area of $50cm^2$ at 30keV.

Figure 4. The first hard X-ray telescope using multilayer supermirror.

Figure 5. Replica foil mirrors made for the first hard X-ray telescope.

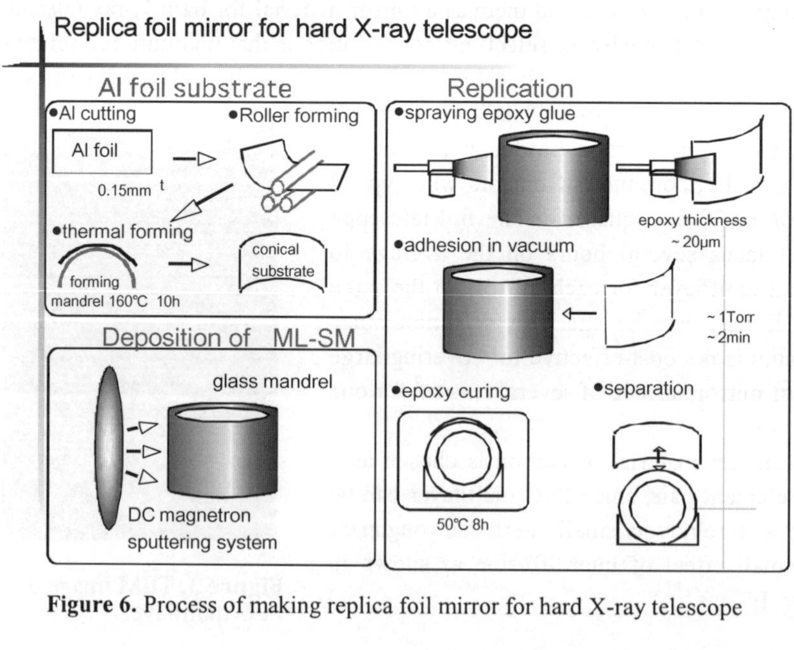

Figure 6. Process of making replica foil mirror for hard X-ray telescope

Since the X-ray telescope uses extreme grazing incidence optics, it is essential to use very thin substrate of X-ray reflector [6] (Fig. 5) to avoid the loss of effective area.

Figure 6 shows the process how to make such thin foil mirrors. We use Al foil as a substrate and epoxy replication method to get smooth and multilayered X-ray reflector [7, 8, 9].

Figure 7. DC magnetron sputtering system

To deposit multilayer on mandrel, we use 2 kinds of DC magnetron sputtering system (DC-2, Fig. 7) and recently introduced Ion Beam Sputtering System (IB-1, Fig. 8). Typical operating condition of these systems is described in table 1.

Ion beam sputtering system is mainly used to study the improvement of multilayer quality, that is, to get smaller interfacial microroughness. To make X-ray reflector, we still use DC magnetron sputtering system.

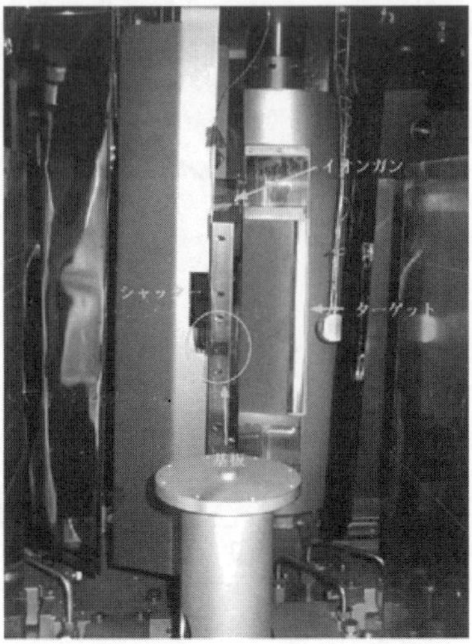

Figure 8. Ion Beam sputtering system

Name		DC-2	IB-1
Sp. Gas / press (Pa)		Ar / 0.4	Ar / 0.02
DC Voltage / Vb (V)		500-900	700-1000
Deposition rate	Pt (A/s)	0.05-0.40	0.1-0.2
	C (A/s)	0.007-0.07	0.03-0.06
Distance (cm)		15	15
Target size (cm)		7.5 x 25	8 x 25

Table 1. Typical operating condition of sputtering system

4. Design of multilayer supermirror

In order to make a hard X-ray telescope with large effective area, we have studied several way of design of multilayer supermirror [10, 11]. After successful fabrication of supermirror which covers X-ray energy up to 40keV, we have been trying to extend energy band up to 80keV.To design such supermirror, we set following constraint. The minimum d-spacing of Pt/C multilayer is 2.4 nm, based on our previous effort to make short period multilayer. From the practical viewpoint we set maximum total bilayer number at 200.

Under these constraints, we optimized multilayer parameters to maximize integrated reflectivity (IR) for given energy band.

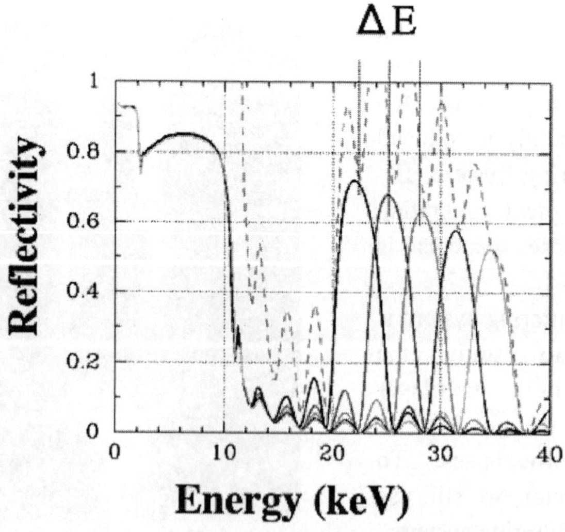

Figure 9. X-ray reflectivity of multilayers with same Bragg peak width.

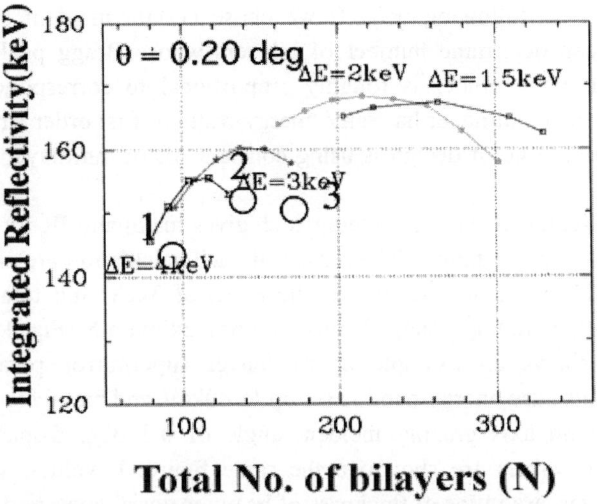

Figure 10. Results of parameter search for multilayer supermirror.

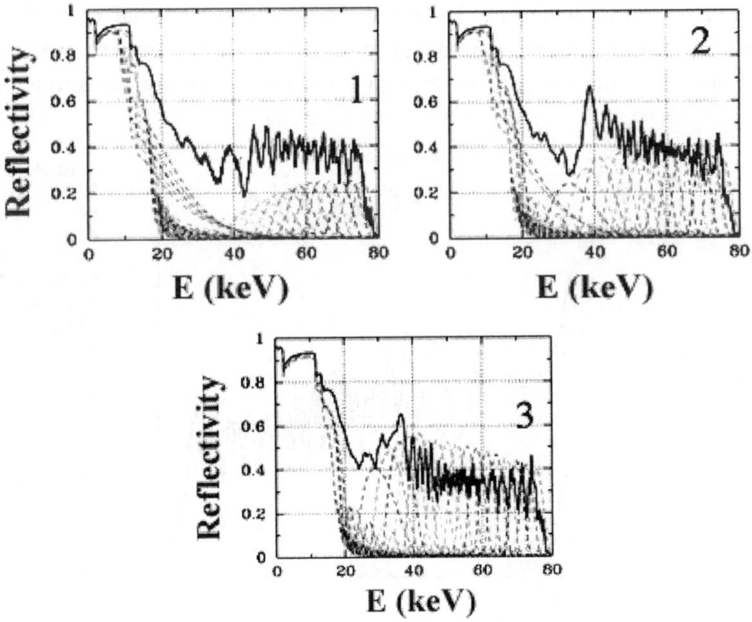

Figure 11. Reflectivity of supermirrors. Numbers in each figure correspond to those in Fig. 10.

If we can neglect absorption effect, it is simple to design as shown in neutron optics. But we cannot do in our case, so we searched for the best parameters in the following way. If we chose certain incident angle and d-spacing, we can determine number of bilayers whose Bragg peak reflectivity saturate. And such number is roughly proportional to corresponding energy, which means each multilayer has same energy width of first order Bragg peak.

So the guide line of design is using constant ΔE of multilayer for stacking (Fig. 9).

Then we searched for parameters which gives maximum IR for given set of ΔE and total bilayer numbers. The results of such calculation are shown in Fig. 10 and Fig. 11. From this result we chose ΔE of 3keV and total number of bilayers as 140, including practical constraint of maximum N (Fig. 10).

Fig. 12 shows an example of multilayer supermirror parameters thus obtained. In this case energy band is set up to 80keV and parameters are for the mirror whose on axis grazing incident angle of 0.2 deg, d-spacing, bilayer number and Γ values are shown in the same figure. Γ values, which is not mentioned so far, are ratios of thickness of heavy material layer to d-spacing and is basically set to 0.4 to maximize the first order Bragg peak reflectivity.

For whole mirrors nested in one telescope, similar multilayer parameters are calculated as shown in Figure 13. It should be mentioned that overall energy response of full telescope can be very smooth, since complex structure of reflectivity as a function of energy can be smeared out by numerous nested

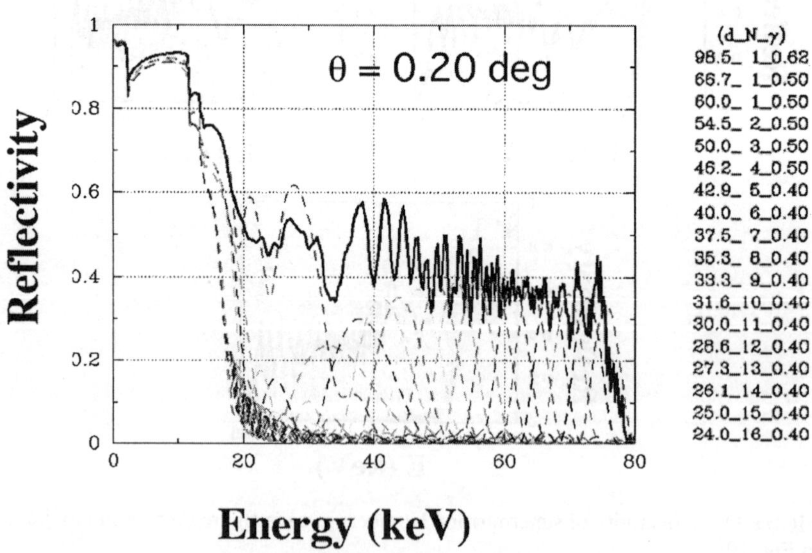

Figure 12. Calculated X-ray reflectivity of supermirror designed for one of the foil mirror in hard X-ray telescope.

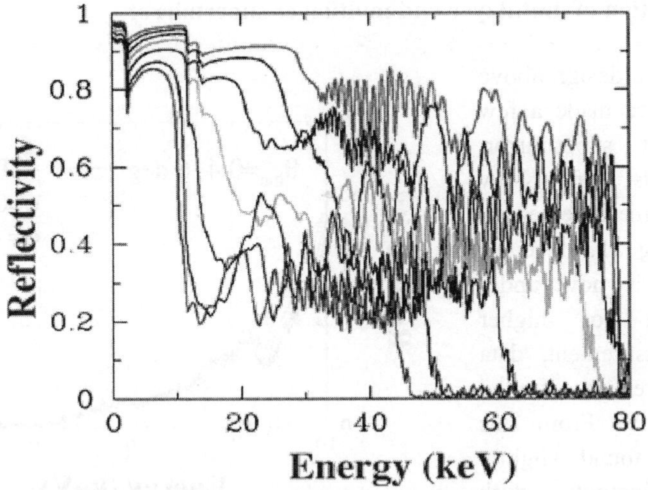

Figure 13. Same as figure 12, but shown for various foil mirrors in one full telescope.

mirrors.

Using this method of optimization, we tried several case which have different maximum energy of observational band as shown in Fig. 14. It is found that we can extend energy band without serious loss of lower band integrated reflectivity.

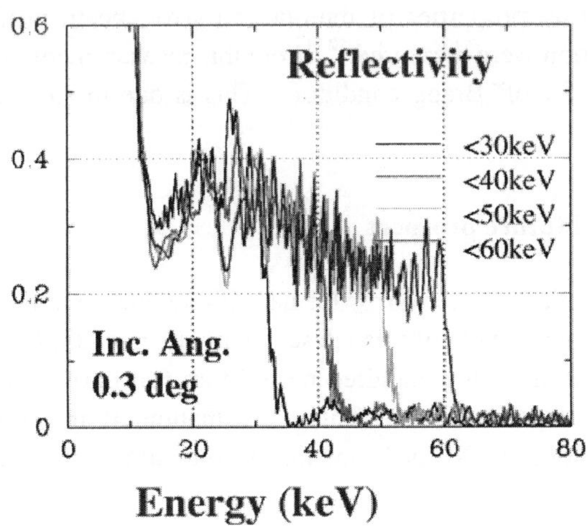

Figure 14. Optimized reflectivity calculated for various maximum energy of observational band.

5. Evaluation of multilayer and multilayer supermirror

Based on the design above described, we made a few samples of supermirror. The results of X-ray reflectivity measurement is shown in Fig. 15. Since our beam line is now under construction for higher energy measurement, data shown here is limited below 50keV. From this figure we found slightly worse reflectivity with interfacial roughness of 4.3Å in terms of Debye-Waller factor.

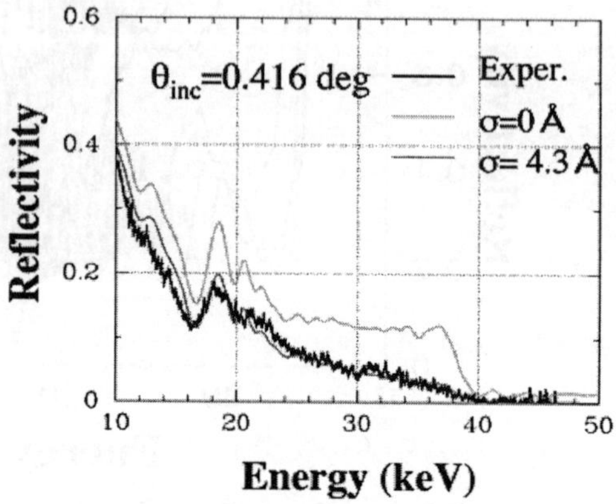

Figure 15. X-ray reflectivity of sample supermirror designed for X-ray telescope.

To Study basic properties of multilayer, we use synchrotron facility of SPRING8, where we can use narrow and high intensity X-ray beam. On the quality of multilayer, it was found that one of the best sample multilayer (d-spacing of 37.4Å, total bilayer number of 20, Γ of 0.44) showed highest reflectivity of 71% at the first Bragg peak for 32keV and a minimum interfacial roughness of 0.285Å.

On the scattering properties of multilayer mirror, beam profiles near 1-st order Bragg condition were measured [12]. From this measurement, we can see that non-specular peak for off Bragg conditions. This is due to so called correlated roughness [13, 14].

6. Summary and future prospect

We have made multilayer supermirror for hard X-ray telescope for the first time. The diameter and focal length of this telescope are 40cm and 8m respectively. In this telescope, there are a few hundred nested thin foil mirrors, on which Pt/C depth graded multilayers were deposited. Evaluation of the X-ray reflection performance was done by X-ray beam line in our laboratory and synchrotron facility SPRING8.

We are now trying to make similar telescope but having wider energy band for the next balloon experiment, and we established the design method of multilayer supermirror having energy band up to 80keV.

For Future X-ray satellite mission, such as NeXT, XEUS, this new technology of multilayer supermirror is expected to open totally new hard X-ray universe to us.

Acknowledgments

We thank Prof. N. Ohnishi for providing TEM data of Pt/C multilayer supermirror.

This work was supported in part by a Grant-in-Aid for Scientific Research on Specially Promoted Research, contract No. 07102007, from the Ministry Education, Science, Sports and Culture, Japan.

References

1. Ogasaka, Y., et. al.: *Proc. SPIE* 2000; **4012**: 294.
2. Okajima, T., et. al.: Appl. Opt. 2002; in press.
3. Owens, S., et. al.: *Proc. SPIE* 2000; **4012**: 619.
4. Tawara Y., Yamashita K., Kunieda H., et. al.: *Proc. SPIE* 1996; **2805**: 236.
5. Yamashita K., et. al.: *Appl. Opt.* 1998; **37**: 8067.
6. Serlemitsos P.J, Jalota L., Soong Y., et. al.: *Publ. Astron. Soc. Japan* 1995; **47**: 105.
7. Kunieda, H., et. al.: Appl. Opt. 2001; **40**: 553.
8. Shibata R., Ishida M., Honda H., et. al.: *Proc. SPIE* 1998; **3444**: 598-609.
9. Soong Y., et. al.: *Proc. SPIE* 1995; **2515**: 64.
10. Tamura K., Yamashita K., Tawara Y., et. al.: *Proc. SPIE* 2000; **4138**: 111-119.
11. Tamura K., Kito H., Yamashita K., et. al.: *Proc. SPIE* 2001; **4501**: 152.
12. Okajima T., Ichimaru S., Tamura K., et. al.: *Proc. SPIE* 2001; **4145**: 53.
13. Ghose S.K. and Dev B.N.: *Physical Review B* 2001; **63**: 245409.
14. Lodha, G.S., et. al.: Jpn. J. Appl. Phys. 1999; **38, Suppl. 38-1**: 289.

PURE NUCLEAR REFLECTION FROM $^{natural}FeN_{0.7}/^{57}FeN_{0.7}$ ISOTOPIC MULTILAYER

AJAY GUPTA, MUKUL GUPTA[+], B. A. DASANNACHARYA,
Inter -University Consortium for DAE facilities, Khandwa Road, Indore-452 017, India.
[+]LNS ETHZ & PSI, CH-5232 Villigen PSI, Switzerland.

Y. YODA, S. KIKUTA
JSAERI, 1-1-1 Kouto Mikazuki-Cho Sayo-gun Hyogo 679-5198, Japan.

M. SETO
RRI, Kyoto University, Kumatori, Sennan, Osaka 590-0494, Japan

Isotopic multilayer of $^{natural}FeN_{0.7}$ /$^{57}FeN_{0.7}$ has been prepared using reactive ion beam sputtering. SIMS measurements show that self-diffusion of Fe in the film is low and is typical of amorphous materials. The periodic multilayer of composition: Float Glass (substrate)/^{nat}FeN(30Å)/[57FeN(20Å)/^{nat}FeN(30Å)]$_{x10}$ was used for measurement of nuclear resonance reflectivity at the NRS beamline of SPring-8 synchrotron radiation source. Nuclear resonance reflectivity, which is delayed in time, was separated from prompt electronic reflectivity using time filtering. A pure nuclear Bragg peak because of isotopic periodicity was observed at an angle of 0.55 deg. (for 14.4 KeV radiation) where electronic scattering was of the order of 10^{-3}. Use of nuclear resonance reflectivity in the measurements of self-diffusion of Fe in the compound is discussed.

1. Introduction

Considerable interest persists in developing nuclear resonant monochromators for synchrotron radiation down to bandwidth of 10^{-9}-10^{-6} eV. Pure nuclear diffraction from antiferromagnetic single crystals was observed as early as in 1978[1]. Use of such crystals in achieving monochromatisation down to10^{-8} eV has been amply demonstrated [2]. However such monochromators suffer from some drawbacks: (i) The angular acceptance of such monochromators are ~20μrad which is small as compared to the divergence of synchrotron radiation. This leads to considerable loss of intensity. (ii) The diffracted radiation from such crystals has a complicated energy spectrum because of antiferromagnetic nature of the crystals, which is an essential requirement for achieving pure nuclear reflection. Grazing incidence antireflection (GIAR) films have also been proposed for suppression of electronic reflectivity [3,4]. These are bilayer structures having one of the layers as resonant Mössbauer nuclei. The layer thickness and density of non-resonant layer is chosen in such a way that reflection from upper and lower interfaces are equal in magnitude and 180^0 out of phase for any non-resonant radiation, where as at resonance this condition is destroyed and nuclear interaction gives strong nuclear scattering. However these films operate at very small angle of incidence typically 4-5 mrad and have small angular acceptance. Also the suppression of electronic scattering with such structure is about 10^{-2}.

On the other hand monochromators based on artificial synthetic multilayers are more promising as they have the advantages that angular acceptance, hyperfine interactions, Bragg angle and angular width of reflection can be tailored easily by varying the structure of multilayer. These are essentially alternate layers of differing isotopes of an element with one of the isotopes being Mössbauer active (e.g. ^{57}Fe). Such structure will act as periodic multilayer only for the radiation in resonance with the Mössbauer transitions of the nucleus, where as off-resonance the entire media works like a single layer. Therefore a Bragg peak is expected from such multilayer having purely nuclear origin. Several attempts have been made to achieve pure nuclear Bragg reflections from nuclear multilayers. Earlier attempts to prepare alternate layers $^{56}Fe/^{57}Fe$ were not very successful because of large interdiffusion at the interfaces. Chumakov et al [5] introduced a thin layer of Scandium between adjacent layers of ^{56}Fe and ^{57}Fe in order to arrest the interdiffusion. Multilayer having structure [^{57}Fe(22Å) / Sc(11Å) / ^{56}Fe(22Å) / Sc(11Å)]$_{x25}$ gave electronic Bragg peak due to electronic charge scattering at double the angle than that of nuclear Bragg peak. However there was significant electronic reflectivity at the nuclear Bragg peak in their spectrum. Röhlsberger et al [6] used an amorphous alloy of Fe_5B_4C as layer material. Existence of pure nuclear Bragg peak was demonstrated using a ^{57}Co radioactive lab source. More recently Deák et al [7] have succeeded in preparing the multilayer of $^{56}Fe/^{57}Fe$ without significant interdiffusion at the interface using MBE technique. Synchrotron radiation has been used to measure the prompt electronic and delayed resonant reflectivity using a structure ^{57}Fe(22.5Å)/^{56}Fe(22.5Å)]$_{x15.}$

In the present work observation of pure nuclear reflections from a synthetic multilayer of structure Float Glass (substrate)/^{nat}FeN (28.5Å)[^{57}FeN(19.5Å)/ ^{nat}FeN (28.5Å)]$_{x10}$ has been presented to explore the possibilities of using such multilayer as Nuclear Bragg monochromator.

2. Theoretical considerations

At grazing incidence geometry the index of refraction of any material is generally given by:

$$n = \left(1 + \frac{\lambda^2}{\pi} \sum_{\mathrm{i}} \rho_i f_{0\mathrm{i}}\right)^{1/2} = 1 + \frac{\lambda^2}{2\pi} \sum_{i} \rho_i f_{0i}^{e} \qquad (1)$$

Where f_0^e is electronic scattering amplitude, ρ_i is density of atom of type i, and λ is the wavelength of the radiation used. The electronic scattering amplitude is given by:

$$f_0^e = -(Z + \Delta f')r_e + \frac{i\sigma_e}{2\pi\lambda} \tag{2}$$

Where $r_e = e^2 / mc^2$, is the classical radius of electron, and σ_e is the total absorption cross section. The first term Zr_e is due to Thomson scattering and gives dominant real contribution. The second term, $\Delta f'r_e$ is due to anomalous scattering and is appreciable near or below an absorption edge. The imaginary term $i\sigma_e / 2\pi\lambda$ represents the absorption of X-rays in the material.

In case, the energy of the incident photon is close to the nuclear transition energy, nuclear resonance scattering can occur and under this condition the total scattering amplitude is given as a sum of two terms:

$$f_0 = f_{0i}^e + f_{0i}^n \tag{3}$$

where f_{0i}^n is the nuclear resonant scattering amplitude. In contrast to the electronic scattering amplitude, nuclear resonant scattering amplitude leads to strong polarisation mixing and thus can be written as 2×2 matrix with its element given by [3]:

$$\left(f_0^N\right)_{ab} = 2\lambda P f_{LM}\left[\frac{\Gamma_\gamma}{\Gamma}\right]\sum_{M=-L}^{L}\sum_{m_0}^{j_0} C^2(j_0 L j_1; m_0 M)\hat{\varepsilon}_a^* \bullet Y_{LM}^{(\lambda)}(\hat{k}_0)\left[Y_{LM}^{(\lambda)}(\hat{k}_0)\right]^* \bullet \hat{\varepsilon}_b / [X(m_0 M) - i)] \tag{4}$$

where Γ_γ is the radiative width (due to radiative, internal conversion and inhomogeneous broadening), and

$$X(m_0 M) = 2\left[E(j_1, m_0 + M) - E(j_0, m_0) - \eta\omega\right] / \Gamma_0; \quad \text{with } M = \Delta J_z = m_1 - m_0 \tag{5}$$

here P is the enrichment of resonant nuclei, f_{LM} is Lamb-Mössbauer factor of the Mössbauer isotope in the sample, α is the internal conversion coefficient (α= 8.21 for ^{57}Fe), J_i are the spin quantum numbers of ground and exited states respectively. (λ)=0 or 1 designates the multipole [$(L,1) = EL$ = electric 2^L pole; $(L,0)=ML$= magnetic 2^L pole] and $Y_{LM}^{(\lambda)}(\hat{k}_0)$ is the vector spherical harmonic which gives the polarization of the radiation emitted in $\hat{k}_0$ direction by an Lλ oscillator with $\Delta J_z = M$. For M1 transition (L=1, λ=0 corresponding Mössbauer transition in ^{57}Fe) expression for $Y_{LM}^{(\lambda)}$ is:

$$Y_{10}^{(0)} = i(3/8\pi)^{1/2}(\sin\theta)\hat{e}_\phi, \qquad Y_{1\,\pm1}^{(0)} = (3/16\pi)^{1/2} e^{\pm i\phi}[\hat{e}_\theta \pm i(\cos\theta)\hat{e}_\phi] \tag{6}$$

Where θ and ϕ are the polar and axial angles specifying the photon direction and quantisation axis. Therefore in general the nuclear scattering amplitude will depend upon the relative direction between the γ-ray direction and the internal field. In the simplest case when there is no Zeeman splitting or quadrupole, the nuclear scattering amplitude simplifies to $(f_0^N)_{ab} = \delta_{ab} f_0^N$, where

$$f_0^N = \frac{2}{\pi} \frac{\lambda^3 f_{LM}(2j_1+1)}{(1+\alpha)(2j_2+1)} \frac{\mathrm{A}}{\mathrm{X - i}} \tag{7}$$

Where $X = 2\Delta E / \Gamma_0$; ΔE denotes the deviation of the photon energy from the resonance energy. The factor A (*Γ=AΓ₀*) denotes the inhomogeneous broadening of the resonance line.

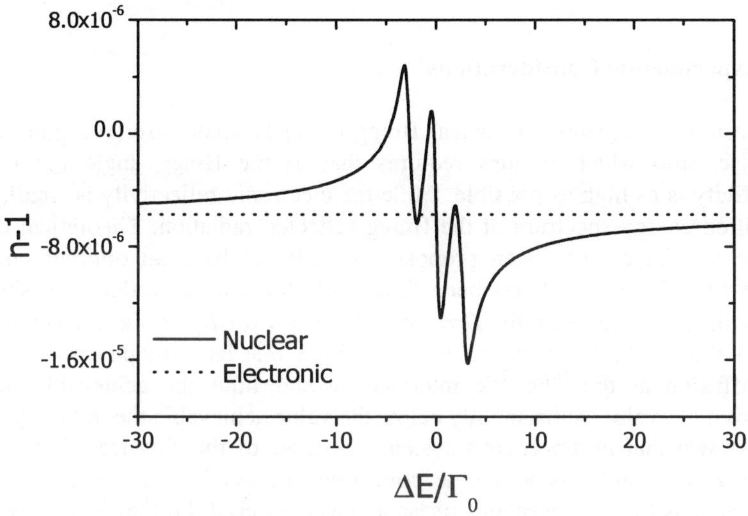

Figure 1: Variation of refractive index with energy for $FeN_{0.7}$; the solid line represents the nuclear term which varies strongly near the nuclear resonance energy, while the dotted line represents the electronic contribution which is almost constant over the energy range of interest.

In equation (3), while electronic scattering factor involves summation over all the different types of atoms in the sample (both Fe and N atoms in the present case), the nuclear resonant scattering factor involves only Mössbauer active nucleus (^{57}Fe). However in case the iron atoms exist in more than one type of environments the weighted sum has to be done over all possible environments. Since the width of nuclear resonance is quite sharp, the nuclear scattering amplitude strongly depends on the photon energy. Figure 1 gives the calculated refractive index of the $^{57}FeN_{0.7}$ layer using the known Mössbauer spectra of the same (Figure 2). Because of the term *1/(X-i)* in Eq.(7) the nuclear resonance factor is appreciable only close to the nuclear transition energy. Therefore in an isotopic multilayer, the X-ray scattering contrast between adjacent layers would exists only for photons having energy close to nuclear transition energy. And therefore, only those photons would contribute to the Bragg peak corresponding to the periodicity of the multilayer.

The total reflectivity is determined by using the expression (1), (3) and (7) for the refractive index and averaging over the energy spectrum of the incident radiation. The nuclear reflectivity is determined by subtracting the electronic part from the total reflectivity. Experimentally the nuclear reflectivity is obtained by time filtering of the reflected radiation. The electronic reflectivity, which is prompt, occurs within few ns of the incident pulse and is blocked electronically [10].

3. Experimental Considerations

The basic requirements for nuclear Bragg monochromator are (i) a good signal to noise ratio which in turn requires that at the Bragg angle the nuclear reflectivity is as high as possible, while the electronic reflectivity is small, (ii) a simplified energy spectrum of the Bragg reflected radiation. Theoretical design for the multilayers has been proposed in order to have an optimum nuclear reflectivity [7,8,9,10]. It has been found that the nuclear reflectivity shows a maximum at Bragg peak for $\beta = d_{57}/d$ (where $d = d_{56} + d_{57}$, is the bilayer period) around 0.3 to 0.5[7]. Experimentally a finite interface roughness as well as interdiffusion at the $^{56}Fe/^{57}Fe$ interfaces would limit the achievable nuclear reflectivity to values substantially below the value achievable theoretically. It has been known that in amorphous systems, because of the absence of grains, the film surface should be smoother as compared to crystalline structures. Thus it is advantageous to use amorphous phase as layer material. Further it is known that in amorphous alloys the self-diffusion of atoms is significantly lower as compared to that in their crystalline counterpart. This happens because of the fact that while in crystalline systems atomic diffusion takes place through vacancy assisted jumps of individual atoms, in amorphous alloys atomic diffusion is more similar to liquid state where collective motion of a group of atoms is involved [11,12,13,14]. The second requirement of having a simplified energy spectrum of Bragg reflected radiation means that the layer material should be non-magnetic. The above requirements suggest that an amorphous and non-magnetic alloy of iron should be best suited as layer material in multilayer monochromator. Further in order to have small non-resonant absorption of the material the alloying element should have low Z. Considering the above requirements possibility of using an amorphous nitride of iron as layer candidate was explored. Earlier studies have shown that iron nitride phases having more than 33% nitrogen are non-magnetic in nature. Therefore in the present work, the techniques of Pulsed Laser Deposition (PLD) and reactive ion beam sputtering (IBS) were tried for depositing the non-magnetic nitride of iron. However it was found that in the case of PLD, only crystalline phases were formed with very high surface roughness of the film [15]. On the other hand, using reactive IBS an amorphous and non-magnetic nitride of iron with composition $FeN_{0.7}$ could be formed [16]. This amorphous phase was used as layer material for further studies.

A beam of nitrogen ions (~700 eV, 90(±1) mA) obtained from a 3 cm broad-beam Kaufman-type hot-cathode ion-gun (Commonwealth Scientific Corporation) was used to sputter iron targets. A base vacuum of $1x10^{-7}$ Torr was achieved before deposition. The nitrogen gas (of purity 99.9995 %) was introduced in the ion source at 7.9 cubic cm/ min using a mass flow controller (MKS-MFC 1179A). The deposition was carried out at room temperature. The targets mounted on a rotary motion feedthrough were kept at an angle of about 45^0 with respect to the incident ion beam and the substrates were kept at a distance of about 120 mm from the target in a direction normal to the ion beam. By switching alternately between ^{nat}Fe and ^{57}Fe targets the isotopic multilayer was deposited on smooth float glass substrate of size 70x30 mm^2 with thickness 3 mm. Considering the theoretical aspects discussed above parameter β was chosen to be 0.4, while the bilayer thickness was taken to be 50 Å, with 10 number of bilayers. Therefore Bragg peak is expected at 0.53 deg. for 14.4 KeV radiation where electronic reflectivity is of the order of 10^{-3}.

4. Results and Discussion

X-ray Photoelectron Spectroscopy (XPS) measurements on the multilayer were done to determine the stoichiometry of the multilayer. After normalizing the areas under the peaks corresponding to iron and nitrogen (Fe-2$p_{3/2}$ and N-*1s* levels) by their cross sections, stoichiometric ratio of iron and nitrogen was found to be 1: 0.7, which indicates that the stoichiometry of the multilayer is $FeN_{0.7}$ [16]. The conversion electron Mössbauer spectrum (CEMS) of the multilayer obtained using ^{57}Co source in Rhodium matrix is shown in Figure 2. Comparison of the Mössbauer spectrum of the present sample with those of a series of metastable iron nitrides produced by *rf* sputtering [17] also suggests the stoichiometry of the sample is close to $FeN_{0.7}$. The spectrum shows nearly a broad singlet in the Mössbauer spectrum, which ensures a simplified energy spectrum of the Bragg reflected radiation in nuclear resonance reflectivity of synchrotron radiation.

The X-ray diffraction pattern of the film shows a broad hump around 2θ=38.7^0 characteristic of amorphous phases [18]. This corresponds to an average Fe-Fe nearest neighbor distance of $a=1.23\lambda/2\sin\theta$ =2.86Å [19]. It may be noted that generally in transition metal-metalloid (TM-M) glasses, having composition around eutectic ($TM_{80}M_{20}$), the broad hump appears around 2θ=44^0-45^0. A larger Fe-Fe nearest neighbor distance in the present case may be attributed to a larger Nitrogen content.

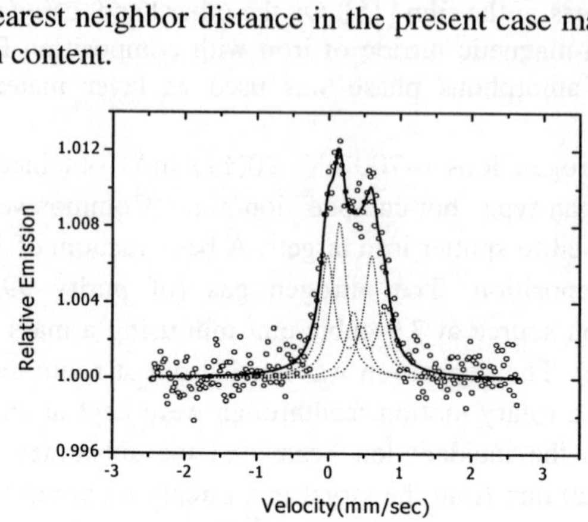

Figure 2: Conversion electron Mössbauer (CEM) spectrum of the $FeN_{0.7}$ isotopic multilayer.

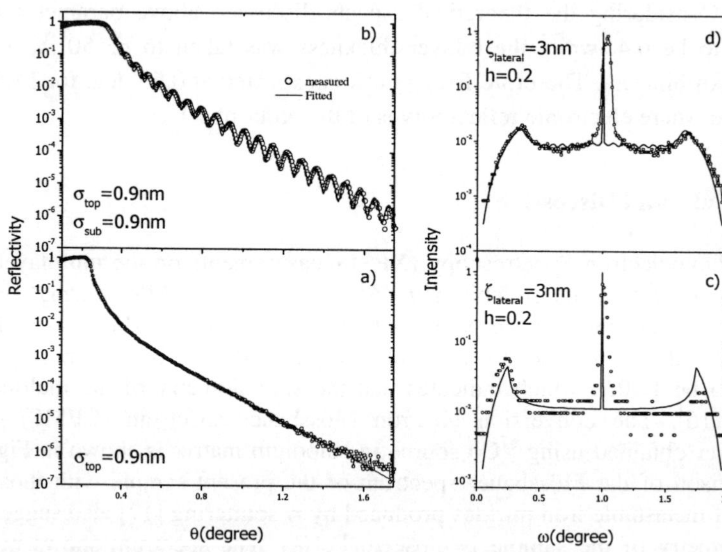

Figure 3: X-ray reflectivity pattern (a) and X-ray diffuse scattering pattern (c) of the bare substrate and that of the $FeN_{0.7}$ isotopic multilayer of the multilayer (b) and (d). It can be seen that the surface roughness, in plane correlation length and jaggedness of the bare substrate and of the isotopic $FeN_{0.7}$ multilayer remains similar.

Figure 3 shows the specular X-ray reflectivity and X-ray diffuse scattering pattern of the bare substrate as well as that of the isotopic multilayer. The measurement was done using Siemens-D5000 diffractometer with Cu-Kα x-rays in the asymmetric Bragg-Brentano geometry with thin film attachment. The reflectivity pattern was fitted using a computer program based on Parratt's formalism [20]. For the substrate the fitted curve gives the value of surface roughness, equal to 9±0.5Å. In the reflectivity pattern of the multilayer the Kiessig oscillations corresponding to total thickness of the film can be clearly seen, however there is no Bragg peak present in the pattern corresponding to the periodicity of the multilayer. This shows that there is no variation in the electron densities between the $^{nat}FeN_{0.7}$ and $^{57}FeN_{0.7}$ layers. Hence the chemical composition of both the isotopic layers is similar. This is an inherent requirement for such type of nuclear Bragg monochromator, as any periodicity because of electron density variation would give rise to a Bragg peak in the electronic reflectivity, which will contribute to the noise at nuclear Bragg peak. The reflectivity pattern of the multilayer was fitted assuming a single layer. The dispersive part of the refractive index of the multilayer was obtained from the experimental value of critical angle for total reflection, while absorption term was calculated from the known value of electron density and stoichiometry of the multilayer. The values for the refractive index and surface roughness of the substrate were taken as obtained by fitting the reflectivity pattern of the bare substrate. The layer thickness and roughness of the top layer were taken as the only parameters for fitting the reflectivity pattern of the multilayer. The fitted pattern yields the total thickness equal to (508.5±1) Å, as compared to the nominal thickness of 530Å. Therefore the actual composition of the multilayer is: Float Glass (substrate)/ ^{nat}FeN (28.5Å) [^{57}FeN(19.5Å)/ ^{nat}FeN (28.5Å)]$_{x10}$. The roughness of the top surface of the multilayer is 9±0.5Å, which indicated that surface roughness of the film does not increase beyond that of the substrate. Thus, as expected the amorphous nature of the layer material results in smoother interfaces.

Self-diffusion of iron in the system was measured using secondary ion mass spectrometry (SIMS) depth profiling. For this purpose a film of amorphous $FeN_{0.7}$ with two thin marker layers of $^{57}FeN_{0.7}$ with thicknesses of 33Å were deposited. Broadening of the marker layers as a function of annealing temperature and time was measured using SIMS depth profiling, with a typical depth resolution of 40Å. The analysis of SIMS data yielded the activation energy q and the pre-exponent factor D_0 for the self diffusion of Fe as $q= 1.3\pm0.2eV$ and $D_0=exp(-16.5\pm2)m^2/s$[18]. From the correlation between the pre-exponential factor D_0 and activation energy q it has been found that the diffusion in the present system is similar to that found in conventional metallic glasses [18].

Nuclear resonance reflectivity measurements on the isotopic multilayer were done at NRS beamline BL09XU of SPring-8 synchrotron radiation source. X-rays from a standard in-vacuum undulator were monochromatized by a double-crystal pre-monochromator, which is tunable above 9 KeV with the resolution ($\Delta E/E$) of approximately 10^{-4}. Further monochromatisation to a value of about 4 meV is obtained by using a nested channel-cut monochromator, which prepares the incident beam for nuclear resonant scattering. The incident beam was tuned to the nuclear resonance energy by observing the fluorescence from the sample at 90 degrees from the incident beam. A sharp peak in the fluorescence yield corresponds to the zero phonon transitions and signifies the resonance between incident radiation and the Mössbauer levels. The broad background around the zero phonon peak is due to inelastic nuclear resonance scattering (Figure 4)[21] involving creation and annihilation of one or more phonons. Area under the broad background after correcting for multi phonon transitions was used for estimating recoilless fraction in the sample, which was obtained to be 0.7.

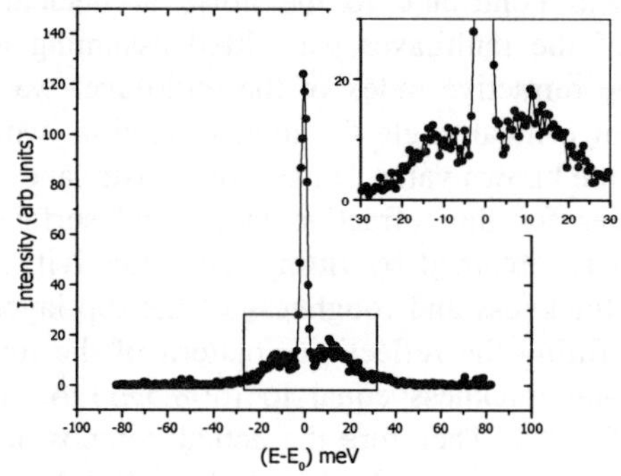

Figure 4: Energy spectrum of nuclear resonant scattering from the $FeN_{0.7}$ / $^{57}FeN_{0.7}$ isotopic multilayer.

This experimentally obtained value of recoilless fraction was used as input parameter for fitting the nuclear resonance reflectivity data, as described in the later part of this section. For the nuclear resonance reflectivity measurement the incident radiation tuned to the nuclear transitions energy and the reflected intensity was measured as a function of scattering angle under specular conditions. The detector used in the experiment was an APD, which has a time resolution of about 1ns. Nuclear resonance reflectivity, which is delayed in time, was separated from the prompt electronic reflectivity by counting the photons in the time window from 40 ns to 155 ns relative to the incidence pulse of photons. Figure 5a shows the nuclear resonant reflectivity pattern from the isotopic multilayer. At resonance energy the nuclear scattering factor associated with ^{57}Fe nucleus gives rise to a Bragg peak at $\theta = 0.55°$. The nuclear reflectivity at the Bragg peak is about 3% where as the electronic reflectivity at this angle is of the order of 10^{-3}, giving a good signal to noise ratio. The nuclear reflectivity pattern was fitted by modifying the formalism for X-Ray reflectivity by adding the structure factor due to nuclear interaction in the total structure factor and averaging over the energy width of the incident radiation.

A program has been developed following the equation (7) to fit the nuclear reflectivity pattern in FORTRAN code. The input parameters for fitting the nuclear reflectivity e.g. thickness, surface and interface roughness and optical constants were obtained by fitting the X-ray reflectivity pattern. The enrichment of the ^{57}Fe isotope was determined by secondary ion mass spectrometry (SIMS) and was found to be 50%. The reason that the enrichment of the ^{57}Fe is low is because the size of the ^{57}Fe foil (which was pasted on a natural Fe sheet) was quite small and not covering the whole ion beam during sputtering. The recoilless fraction as discussed above, was obtained experimentally using inelastic nuclear resonance scattering. Therefore the only parameter varied in fitting the nuclear reflectivity is the depth distribution of ^{57}Fe layers. An interdiffusion of 6Å fits the Bragg peak well and the obtained depth profile for top two layers is shown in Figure 6.

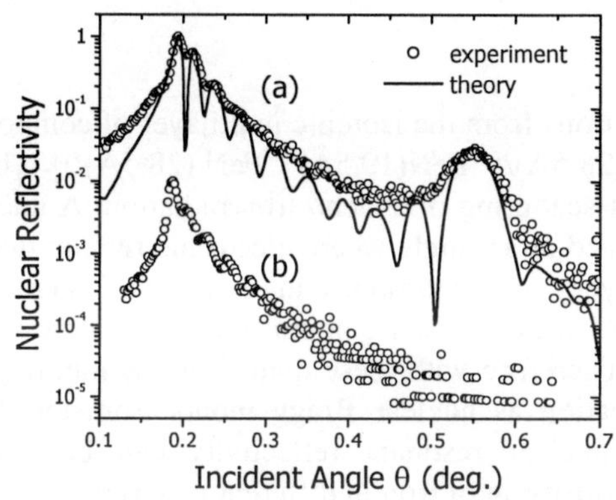

Figure 5: Nuclear resonance reflectivity of the as deposited [Float Glass (substrate)/^{nat}FeN (28.5Å)[^{57}FeN(19.5Å)/ ^{nat}FeN (28.5Å)] $_{x10}$] multilayer a) and after annealing at 250 ^{0}C in vacuum for 40 minute b). The solid line in a) corresponds to fitted curve.

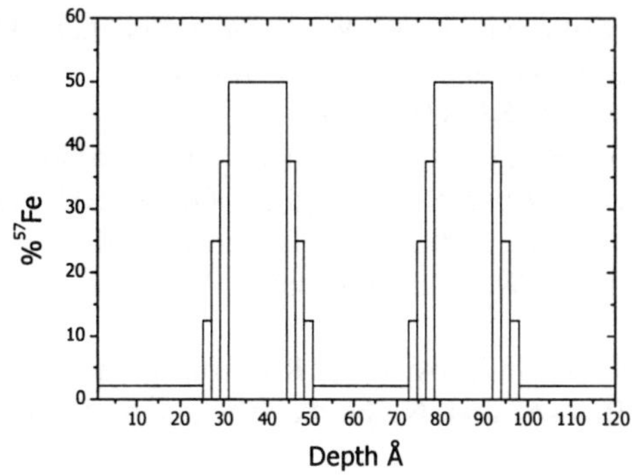

Figure 6: Typical depth-distribution of 57Fe as obtained by the fitting of the nuclear resonance reflectivity data.

Figure 5 (b) gives the nuclear resonance reflectivity of the same multilayer after annealing at 250°C for 40 min. One can see while the total thickness oscillations are still present, the Bragg peak has totally disappeared. This suggests that the above annealing results in an almost complete homogenization of the multilayer. These results also suggests that the nuclear resonance reflectivity can be effectively used for measuring the self-diffusion in multilayers. The advantage of the reflectivity technique over SIMS depth profiling is its higher resolution. Diffusion depths as small as 1Å can be measured using reflectivity, as compared to a resolution of several 10's of Angstroms in the case of SIMS depth profiling. Neutron reflectivity has also been used for measuring self-diffusion in multilayers [22]. However, nuclear resonance reflectivity has advantage over neutron reflectivity in terms of smaller sample size and faster data acquisition.

5. Conclusion

Pure nuclear reflections from the isotopic multilayer of composition: Float Glass (substrate)/^{nat}FeN (28.5Å)/[^{57}FeN(19.5Å) ^{nat}FeN (28.5Å)/)]$_{x10}$ have been obtained in nuclear resonant scattering of synchrotron radiation. A nuclear reflectivity of 3% has been obtained at an angle where electronic reflectivity is of the order of 10^{-3}. Since the layers are amorphous in nature the interdiffusion across the interfaces is low. Also $FeN_{0.7}$ being nonmagnetic in nature reflects a simplified energy spectrum. Therefore with these qualifications multilayer of such type is suitable for application as nuclear Bragg monochromator. Present study also demonstrate that nuclear resonant reflectivity can be effectively used for measuring the self-diffusion of iron in different materials.

References

1. R. L. Cohen, G. L Miller and K. W. West, *Phys. Rev. Lett.* **41,** 381(1978).
2. E. Gerdau, R. Rüffer, H. Winkler, W. Tolksdorf, C. P. Klages and J. P. Hannon, *Phys. Rev. Lett.* **54**, 835(1985).
3. J.P. Hannon, N. V. Hung, G.T. Trammell, E. Gerdau, M. Mueller, R. Rüffer and H. Winkler, *Phys. Rev.* **B32**(8), 5068 and 5081(1985).
4. R Röhlsberger, E. Gerdau , M. Harsdorff, O. Leupold, E. Lüken, J. Metge, R.Rüffer, H. D. Rüffer, W. Sturhahn, and E. Witthoff, *Europhys. Lett.* **18**, 561(1992).
5. A. I. Chumakov, G. V. Smirnov, A. Q. R.Baron, J. Arthur, D. E. Brown, S. L. Ruby, G. S. Brown, and N. N. Salashehenko, *Phys. Rev. Lett.* **71**, 2489(1993).
6. R Röhlsberger, E. Witthoff, E. Gerdau and E. Lken, *J. Appl. Phys.* **74**,1933(1993).

7. L. Deák, G. Bayreuther, L. Bottyán, E. Gerdau, J.Korecki, E.I.Kornilov, H.J. Lauter, O. Leupold, D.L.Nagy, A.V. Petrenko, V.V. Pasyuk-Lauter, H. Reuther, E. Richter, R. Röhlsberger, E. Szilagyi, *J. Appl. Phys.* **85**(1),1(1999).
8. A. I. Chumakov and G. V. Smirnov, *JETP Lett* **53**, 273(1991).
9. L. Deák, L. Bottyán and D. L. Nagy in Condensed Matter Studies by Nuclear Methods, Proc. XVIII, Zakopane School of Physics, Zakopane, 1993, edited by E.A.Gorlich and K. Tomala (Institute of Nuclear Physics, Kraków, 1993), p.269.
10. L. Deák, L. Bottyán, and D. L.Nagy, *Hyperfine Interactions* **92**,1083(1994).
11. A. Heesemann,V. Zöllmer,K. Rätzke and F. Faupel, *Phys. Rev. Lett.* **84,** 1467 (2000).
12. K. N. Tu and T. C. Chou, *Phys. Rev. Lett.* **61,** 1863 (1988).
13. A. Grandjean, P.Blanchard and Y. Limoge, *Phys. Rev. Lett.* **78,** 697 (1997).
14. P. Klugkist, K. Rätzke, S. Rehders, P. Troche and F. Faupel, *Phys. Rev. Lett.* **80,** 3288 (1998).
15. Mukul Gupta, Ajay Gupta, P. Bhattacharya, P. Misra and L. M. Kukreja, *Journal of Alloys and Compounds* **326**, 265(2001).
16. Mukul Gupta, Ajay Gupta, S. M. Chaudhari, D. M. Phase, V. Ganesan, M. V. Rama Rao, T. Shripathi, B. A. Dasannacharya, *Vacuum* **60,** 399(2001).
17. T. Hinomura, Ph.D. Thesis, Osaka University, Osaka, Japan (1998).
18. Mukul Gupta, Ajay Gupta, S. Rajagopalan, and A. K. Tyagi, *Phys. Rev.* **B 65**, 214204 (2002).
19. A. Guinier, "X-ray Diffraction –In Crystals, Imperfect Crystals and Amorphous Bodies" Dover publications, New York (1994).
20. L. G. Parratt, Phys. Rev. **95,** 359 (1954).
21. M Seto, Y. Yoda, S. Kikuta, X. W. Zhang and M. Ando, *Phys. Rev. Lett.* **74**, 3828(1995).
22. N. Cowlam, J. Speakman, P. Rose, J. A. Hunt, R.E. Somekh and A.L. Greer, *J. Magn. Magn. Mater.* **156**, 411(1996).

SURFACE MODIFICATION BY ENERGETIC ION BEAMS

SCANNING PROBE STUDIES OF SWIFT HEAVY ION IRRADIATED SEMICONDUCTOR SURFACES

J. P. SINGH[c] AND D. KANJILAL[d]

Nuclear Science Centre, Aruna Asaf Ali Marg, New Delhi 110 067, India

The electronic excitations (S_e) induced surface modifications in n-Si(111) and semi-insulating InP(111) single crystals were *ex situ* examined by atomic force microscopy. The different types of surface modifications were observed in both the materials when irradiated by 200 *MeV* Ag^{+14} and 100 *MeV* Au^{+8} ions. For the S_e insensitive materials such as Si(111), the morphological changes were governed by the cumulative effects of ion beam and results in the formation of nanosize dike structures, whereas for S_e sensitive materials such as InP(111), the effects of individual ions in terms of the formation of nanoscale pit structures were visible. The nuclear energy loss S_n was found to have a decisive role in determining the damage caused by the ion irradiation on InP surface.

1. Introduction

Swift heavy ion (SHI) passes through the solid surface with a velocity comparable to the Bohr velocity of electron and loses its energy while traversing through the matter. The total energy loss can be expressed mainly as the summation of two energy loss processes. They are electronic energy loss (S_e) due to the inelastic collisions with electrons and nuclear energy loss (S_n) due to the elastic collisions with atoms of the solid with the projectile ion. The modification of semiconductor surfaces by SHI beam has been studied by several researchers [1-3] and it is observed from the different experiments that SHI induced modifications on semiconductor surfaces depends on whether it is elemental or compound semiconductor. The earlier experiments [4] involving giga electron volts (GeV) energies have proved that it is not possible to create amorphized latent tracks in bulk silicon by mono-atomic ions (such as Au, Ag, I etc.). There is a threshold value $(S_e)_{th}$ (~30 *keV/nm*) of the electronic energy loss S_e for the creation of amorphized latent tracks in the crystalline Si [5], which is normally not achievable by mono-atomic MeV ion beam. Therefore, Si is generally termed as S_e insensitive materials. Whereas, Herre *et al.* [6] observed the formation of tracks in InP after irradiation by 250 *MeV* Xe ions at

[c] Present address: Department of Physics, Applied Physics and Astronomy, Rensselaer Polytechnic Institute, Troy 12180-3590 NY

[d] Corresponding author E-mail: dk@nsc.ernet.in

room temperature. The $(S_e)_{th}$ was estimated as 13 *keV/nm*, which is small compared to the threshold S_e of crystalline Si.

The damaged zone produced by SHI is localized within a diameter of a few *nm* along the ion trajectory. Therefore, to study the ion-induced damage on the sample surface the size of the observation window should be limited to an area of a few nm^2. In this respect, atomic force microscopy (AFM) and scanning tunneling microscopy (STM) offer an advantage over conventional techniques to study the SHI induced surface modifications [1,7-8]. Using these techniques, windows of several *nm* can be scanned having resolution in the vertical direction of the order of a few Angstroms.

In order to understand the effect of SHI on S_e insensitive and S_e sensitive materials, we have irradiated crystalline n-Si(111) and semi-insulating InP(111) surfaces by 200 *MeV* Ag^{+14} and 100 *MeV* Au^{+8} ions at varying fluences between 10^{12} to 10^{14} *ions* cm^{-2}. Both the surfaces observed behave differently when exposed to SHI. A new type of dike structure was observed on the Si surface and nanoscale defects in the form of pits and hillocks were observed on the InP surface after the ion bombardment. The Si surface modifies as a result of cumulative action of the ion beam, whereas on InP surface the individual ion effect has been seen.

2. Experimental

The samples chosen for ion bombardment were <111> oriented n-type (P doped) Si single crystals and semi-insulating InP surfaces. The 200 *MeV* Ag^{+14} ions and 100 *MeV* Au^{+8} ions were used for the irradiation. The irradiation was performed with 15 *MV* Pelletron accelerator at Nuclear Science Centre [9]. In the case of Si surface the ion beam had an angle of 15° with respect to the surface normal, whereas for InP the ion beam incident normally on the surface. The thickness of all of the samples was 0.5 *mm*. The samples were mounted on a copper target ladder with silver paste giving good thermal conductivity between them. This prevents sample heating during high-energy irradiation. Prior to the irradiation, Si surfaces were etched in NH_4F solution. This chemical treatment of the Si surfaces results in hydrogen passivated trapezoid shaped terraces uniformly distributed over the entire surface. Thanh *et al.* [10] investigated the surface structure and chemistry of Si surfaces after wet chemical etching in HF and NH_4F solutions. The results suggested that hydrogen passivated (111) facets were formed on Si surfaces only after a prolonged etching in NH_4F solution and they remained stable up to an annealing temperature of 650°*C*. In the present case, the observed terraces are having atomically flat surface and smooth facets, whereas other regions are rather rough. These atomically flat trapezoid terraces can be a place to observe the ion beam induced mass transport on the surface. During irradiation the temperature of the Si samples were maintained at 80 *K*. The chamber pressure during irradiation was 7×10^{-7} *mbar* maintained by a cryo-pump to avoid hydrocarbon contamination during irradiation. The samples were irradiated with fluences varying between 10^{12} to 10^{14} *ions* cm^{-2}. The ion flux was 10^{9} *ions* $cm^{-2}\ s^{-1}$. Fluences were determined by integrating the charges of the ions falling on the sample directly by a current integrator. The secondary electrons were suppressed by a negatively biased (-120 *Volt*) cylindrical suppressor assembly, which encloses the sample holder. Ion fluence is estimated within an accuracy of $\pm$ 5%.

In all experiments, one of the samples was left unirradiated on the target ladder and was used as the pristine (reference) sample. This is to ensure that effect is purely due to ion irradiation and is not an artifact. The sample surfaces were then *ex situ* analyzed using AFM from digital instruments (Nanoscope III) by Si_3N_4 tips. All samples were scanned in trace and retrace directions. This was done to identify the possible direction induced scanning artifacts. The AFM calibration was accomplished by imaging a gold micro grid structure on Si surface before each set of the experiments.

3. Results

The ion beam induced formation of dike structure on the chemically grown terraces on Si(111) surface after 200 *MeV* Ag^{+14} ions irradiation at 15° angle with respect to the surface normal for fluence varying from 10^{12} to 10^{14} *ions* cm^{-2} is systematically presented in Figure 1. The direction of the incident ions is also depicted. The dike structure was observed at the boundary of the terrace. All such dikes were formed in the same side of the incident ion beam. The height of dike structure varies from 7.2 *nm* at 10^{12} *ions* cm^{-2} to 21.1 *nm* at 10^{13} *ions* cm^{-2} and to 5.4 *nm* at 10^{14} *ions* cm^{-2}.

The typical AFM images of InP(111) surfaces after 200 *MeV* Ag^{+14} and 100 *MeV* Au^{+8} ions irradiation are shown in Figure 2. It is observed from the figure that pristine surface is smooth and have noise of the order of 0.5 *nm*. The surface morphology after 200 *MeV* Ag^{+14} ions irradiation of InP surface is shown in Figure 2(b). The nano scale pit type defect formed on InP surface after 100 *MeV* Au^{+8} ion irradiation is shown in Figure 2(c). The average diameter of these pit type defects is 24 *nm*. The areal density of the observed

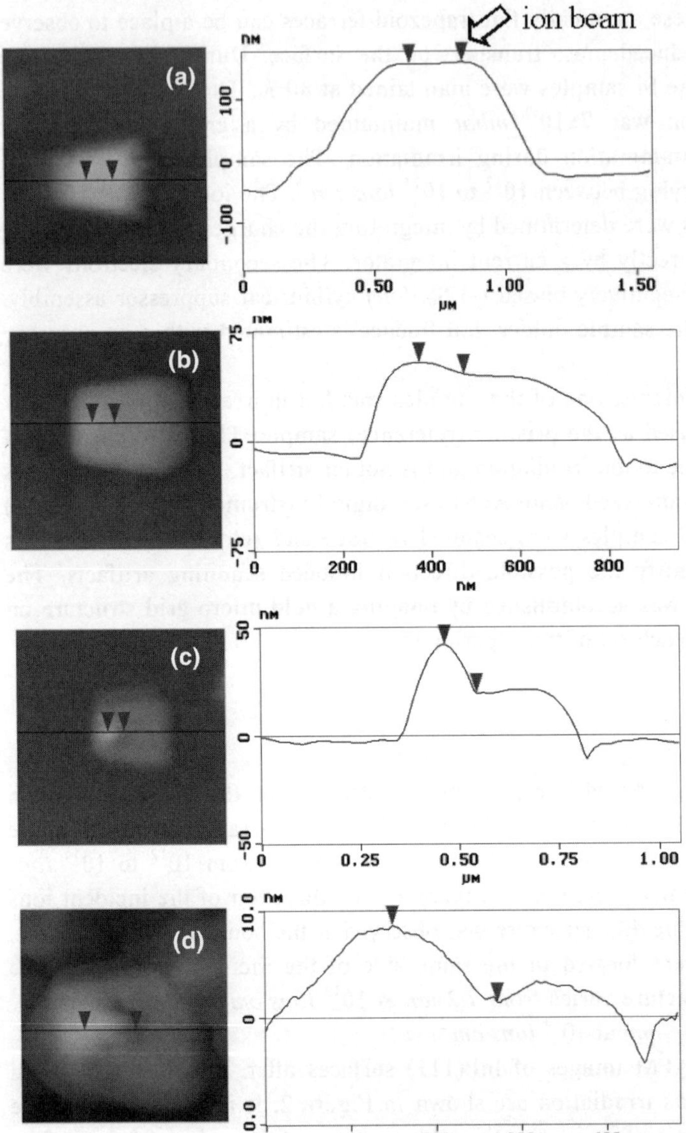

Figure 1. The line profiles of AFM topographies showing the formation of dike structures at the (a) Si(111) terrace induced by 200 MeV Ag^{+14} ions at 15° angle of incidence with respect to the surface normal at the fluences of (b) 10^{12} ions cm^{-2}, (c) 10^{13} ions cm^{-2}, and (d) 10^{14} ions cm^{-2}. The beam direction is also illustrated in the figure.

surface defects was approximately in agreement with the estimated ion fluence incident on the sample surface. So, these defects are single ion impact modifications. Figure 2(d) shows the overlapping of pits after irradiation at the fluence of 10^{14} *ions cm^{-2}*.

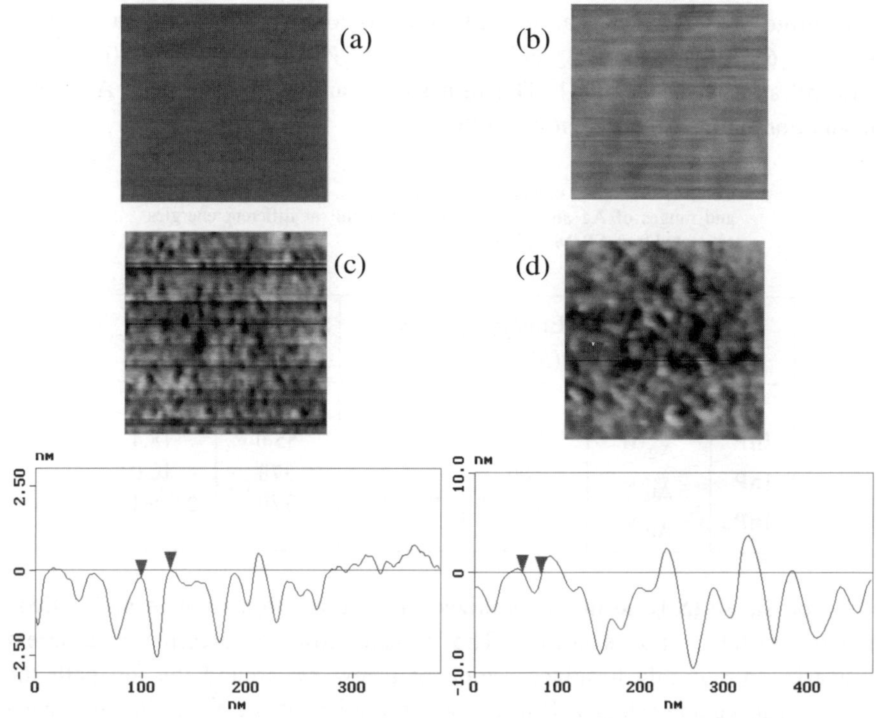

Figure 2. AFM images of InP surfaces (a) pristine, irradiated by (b) 200 *MeV* Ag^{+14} ions at 10^{14} *ions cm^{-2}* fluence and by (c) 100 *MeV* Au^{+8} ions at the fluences of 10^{12} and (d) at 10^{14} *ions cm^{-2}*. Pit-type single-ion impact sites are visible in (c) which are otherwise absent in (b). Images (a) and (b) are of 400×400 *nm^2* area.

4. Discussion

The formation of dike structure on the irradiated Si surfaces as shown in Figure 1 is due to the ion-induced lateral mass transport on the surface. The lateral mass transport phenomenon consists of the displacement of the irradiated material along the surface in the direction of the ion velocity parallel to the surface. This phenomenon occurs only in amorphous material [11-13]. In case of crystalline materials, it becomes apparent only after the material becomes amorphized.

In the case of low fluence regime ($\phi < \phi_c$, where ϕ_c is the critical fluence where the damage zones start overlapping), the damaged regions do not overlap with each other. The total damage increases with fluence linearly. The threshold value $(S_e)_{th}$ of S_e for the creation of amorphized latent tracks in the crystalline Si is about 30 *keV/nm* [5], whereas for InP it was estimated as 13 *keV/nm* [6]. The values of S_e and S_n for 200 *MeV* Ag^{+14} ions irradiation on Si and InP are given in Table 1. The values of S_e and S_n for 100 *MeV* Au^{+8} ions irradiation on InP are also shown in the Table 1.

Table 1. The values of electronic energy loss S_e, nuclear energy loss S_n and ranges of Ag and Au ions in Si and InP at different energies estimated from SRIM 98 [14].

	Ion	Energy (*MeV*)	S_e (*keV/nm*)	S_n (*eV/nm*)	Range (*μm*)
Si	Ag^{+14}	200	12.5	33.5	22.3
InP	Ag^{+14}	200	16.5	55.0	18.4
InP	Au^{+8}	100	15.7	378	10.0
InP	Au^{+8}	550×10^{-6}	39.1×10^{-3}	370	23.0×10^{-4}

For Si, $S_e<(S_e)_{th}$ so the amorphized latent track formation in bulk silicon single crystal is not expected. The surface however presents a different scenario. In a highly localized region of a few *nm* around the ion path the surface can be amorphized. During ion irradiation of the Si surface hydrogen atoms desorb from the passivated Si surface sites. This hydrogen desorption induces surface states and thus leaves unsaturated dangling bonds on the surface. With certain amount of disorder induced by SHI on the Si surface in the form of broken bonds, these active regions become sensitive to the irradiation [15]. Therefore, SHI irradiation in Si may produce surface amorphization without affecting bulk. At around a critical fluence ϕ_c the damaged regions overlap with each other. Therefore, surface gets amorphized and an amorphous/crystalline (a/c) interface is created. The extent of amorphization in a region on the surface is proportional to the number of ions hitting that region. The thickness of the amorphized layer in that region is therefore expected to scale with this multiple-ion hit process. The schematic diagram of SHI irradiation induced amorphization and creation of a/c interface are shown in figure 3.

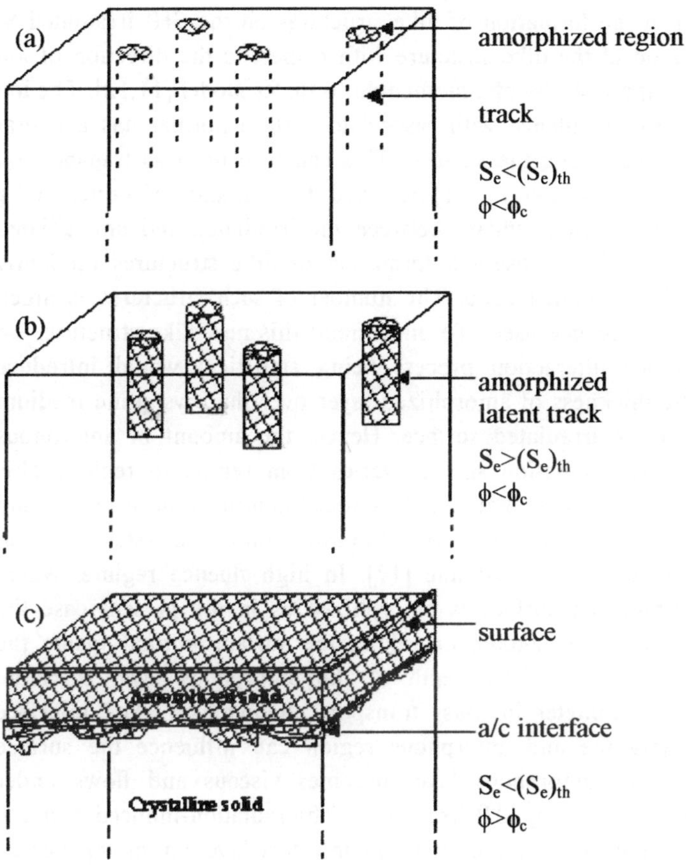

Figure 3. Schematic diagram showing (a) the formation of amorphized surface zone due to the single ion impact for $S_e<(S_e)_{th}$ and (b) amorphized latent tracks for $S_e>(S_e)_{th}$ and (c) the creation of amorphous/crystalline (a/c) interface for $S_e<(S_e)_{th}$ and at the fluence greater than the fluence required for overlapping of the damaged zones.

The amorphous Si material is produced if the energy deposition per unit volume reaches a level of 6×10^{20} *keV cm*$^{-3}$ [16]. The energy density deposited by 200 *MeV* Ag^{+14} ion in Si is 1.61×10^{20} *keV cm*$^{-3}$. So, for the amorphization of Si the number of tracks to be overlap will be about 4. Hence, the ion fluence $\phi\geq 4\phi_c$ is needed for the amorphization and dynamics of a/c interface layer towards the crystalline side. It is also observed experimentally by X-ray diffraction analysis that amorphization starts after the fluence of 5×10^{12} *ions cm*$^{-2}$ and also the dike structure (see Figure 1) becomes prominent only after the fluence of 5×10^{12} *ions cm*$^{-2}$. Thus, it can be argued that a cumulative action of the ion beam is involved in the formation of dike structures on the SHI irradiated Si surface. The location of the dike structure with respect to the direction of ion beam points to the applicability of a momentum transfer model [11,12]. The ion beam at 15° angle of incidence with respect to surface normal has a lateral momentum component along the surface. This can lead to mass transport on the irradiated surface. Cliché *et al.* [11] observed the formation of a ditch and a dike structure at the mask boundary between the irradiated and unirradiated regions. Whereas, we have observed formation of dike structures randomly distributed over the irradiated zone. The number of such structures is much smaller than the ion fluence used. To understand this new dike structure, we note that ion-surface interaction process being statistical would introduce fluctuations in the thickness of amorphized layer over the crystalline medium as measured from the irradiated surface. Hence, the amount of amorphous material beneath the irradiation surface varies from region to region. This situation is very different from that of SHI induced modification in metals and insulators in the low fluence regime, where the evolution of the system with ion fluence is considered only in *xy*-plane [17]. In high fluence regime, where tracks are not formed but surface is amorphized as in the present case the system evolution along the *z*-axis, i.e. along the surface normal inside the material become important. The depth of the amorphous layer therefore becomes a crucial parameter in mass transport phenomenon. The interface layer between crystalline and amorphous region can influence the surface morphology when the amorphous layer becomes viscous and flows under momentum transfer induced by SHI irradiation. The radiation-induced decrease in viscosity reflects the uneven structure of the interface layer on the top surface and hence makes this uneven structure behave like an internal mask. The momentum provided by the incidence ions produce anisotropy in the dike structure with respect to the surface normal. The number of such internal masks determines the number of ditch and dike structures and will be dictated by the probability of multiple-ion hit phenomenon as discussed earlier.

The irradiation of InP surface with 200 *MeV* Ag^{+14} ions is an interesting case. The threshold value $(S_e)_{th}$ for the creation of amorphized latent tracks in the crystalline InP is 13 *keV/nm*, which is less than the S_e of 16.6 *keV/nm* as given in Table 1. So in present case $S_e>(S_e)_{th}$, the amorphized latent track formation in bulk InP single crystal is expected. Thus, in the case of InP, surface as well as bulk is expected to be amorphized. The critical energy density E_c for amorphization (~2.5×10^{20} *keV/cm*^{3}) in case of III-V semiconductor is lower than that of Si (~6×10^{21} *keV cm*$^{-3}$). This is due to the fact that the compound semiconductor lattice contains different atoms in the basis. This makes the contribution in the form of configurational entropy and lowers the free energy. The energy density deposited by 200 *MeV* Ag^{+14} ion on InP surface is ~2.1×10^{20} *keV/cm*^{3} (by assuming 10 *nm* as track diameter) which is of the same order as the E_c for amorphization. Therefore, a single ion impact can create an amorphized latent track. In fact, Herre *et al.* [6] observed the formation of tracks in InP after irradiation of 250 *MeV* Xe ions at room temperature. The nanoscale pit structures observed in the Figure 2 reflect the damaged caused by individual ion bombardment.

It is interesting to compare the values of S_e and S_n as given in Table 1 for irradiation of InP by 200 *MeV* Ag^{+14} and 100 *MeV* Au^{+8}. In both cases of irradiation the S_e values for Au and Ag ions in InP is approximately same, whereas S_n is approximately seven times higher for 100 *MeV* Au ion irradiation than for 200 *MeV* Ag^{+14} ion irradiation. The pit type defects were observed on InP surface after irradiation with 100 *MeV* Au^{+8} ions, which were otherwise absent after 180 *MeV* Ag^{+14} ions irradiation. This indicates that in the formation of nano scale defect on InP surface S_n has a decisive role. To clarify the situation the effect of S_n was separated out from S_e for 100 *MeV* Au^{+8} ion incident on InP. From the Bragg curve of Au ion irradiation in InP, it has been found that 550 *eV* Au ion corresponds to the S_n value as in case of 100 *MeV* Au ion irradiation in InP as shown in Table 1. The value of S_e in this case is approximately ten times smaller than the S_n value. Hence, S_n is a dominant mechanism in this case for defect formation. The lateral extent of the damage due to S_n can be calculated by assuming that the spikes resulting from the displacement cascade are spherical [18], whereas spikes induced by electronic excitations are characterized by a cylindrical geometry. The radius of the spherical distribution (r) will be the range of ion (R_p) ± straggling (s) in the solid. The lateral extent of the damage on InP surface due to S_n was found to be ~6 *nm*, which is much smaller than the experimentally observed pit type defect diameter of 24 *nm*. This shows that the formation of nano scale defect on the surface is not due to S_n solely but is a combined effect of coupled S_e and S_n.

Thus, the electronic excitations may prepare the solid in such a way that large sputtering by elastic collisions occurs, which further results in a larger pit type defect diameter than that created by S_n alone.

5. Conclusions

In summary, we have used atomic force microscopy to reveal the morphological changes on Si and InP surfaces after swift Ag^{+14} and Au^{+8} ions. In case of Si, the nanosize dike structures were observed as the result of lateral mass transport due to the cumulative action of ion beam. For InP surface, nanoscale pit type defects due to individual ion bombardment were observed. Moreover, S_n was found to play a decisive role in the formation of damage on the surface, which is often neglected for S_e dominant processes.

Acknowledgments

Authors are thankful to Dr. V. Ganesan for AFM measurements. We are also thankful to Dr. N.C. Mishra for his fruitful suggestions and discussions. The support received from Department of Science and Technology (DST), India for this work is gratefully acknowledged.

References

1. L.P. Biró, J. Gyulai, K. Havancsák, A. Yu. Didyk, S. Bogen, L. Frey, *Phys. Rev.* **B54**, 11853 (1996).
2. P.I. Gaiduk, E.F. Komarov, W. Wesch, *Nucl. Instr. and Meth.* **B164-165**, 377 (2000).
3. J.P. Singh, R. Singh, N.C. Mishra, V. Ganesan, D. Kanjilal, *Nucl. Inst. and Meth.* **B 179**, 34 (2001).
4. M. Levalois, P. Bogdanski, M. Toulemonde, *Nucl. Inst. and Meth.* **B**63, 14 (1992).
5. B. Canut, N. Bonardi, S.N.M. Ramos, S. Della-Negra, *Nucl. Instr. and Meth.* **B146**, 296 (1998).
6. O. Herre, W. Wesch, E. Wandler, P.I. Gaiduk, E.F. Komarov, S. Klaumunzer, P. Meier, *Phys. Rev.* **B58**, 4832 (1998).
7. J.P. Singh, A. Tripathi, D. Kanjilal, *Vacuum* **57**, 319 (2000).
8. L.P. Biró, J. Gyulai, K. Havancsák, *Vacuum* **50**, 263 (1998).
9. D. Kanjilal, S. Chopra, M.M. Narayanan, I.S. Iyer, V. Jha, R. Joshi, S.K. Datta, *Nucl. Instr. and Meth.* **A238,** 97 (1993).
10. V.L. Thanh, D. Bouchier, G. Hincelin, *J. Appl. Phys.* **87**, 3700 (2000).
11. L. Cliché, S. Roorda, M. Chicoine, R.A. Masut, *Phys. Rev. Lett.*, **75**, 2348 (1995).
12. Chicoine M., Roorda S., Cliché L., Masut R.A. (1997) *Phys. Rev. B* **56**, 1551.
13. S. Klaumuenzer and G. Schumacher, *Phys. Rev. Lett.* **51**, 1987 (1983).

14. J.F. Ziegler, J.P. Biersack, U. Littmark, *Stopping and Ranges of Ions in Matter* (Pergamon, New York) (1985).
15. J.P. Singh, R. Singh, N.C. Mishra, N. Ganesan, D. Kanjilal, *J. Appl. Phys.* **87**, 2762 (2000).
16. G.F. Chiarotti, F. Fumi, M.P. Tosi, *Proceedings of the international school on physics*, Elsevier Science Publisher B.V. , p.532 (1992).
17. Ch. Dufour, E. Paumier, M. Toulemonde, *Nucl. Instr. and Meth.* **B122**, 445 (1997).
18. I.H. Wilson, N.J. Zheng, U. Knipping, I.S.T. Tsong, *Phys. Rev.* **B38**, 8444 (1988).

ION IRRADIATION EFFECTS AND ION BEAM STUDIES OF SEMICONDUCTOR MULTILAYERS

S.V.S. NAGESWARA RAO AND ANAND P. PATHAK[c]
School of Physics, University of Hyderabad, Central University PO, Hyderabad 500 046, India
E-mail: appsp@uohyd.ernet.in

D.K. AVASTHI
Nuclear Science Centre, Post Box No. 10502, Aruna Asaf Ali Marg New Delhi 110 067, India

R. MURALIDHARAN, UMESH TIWARI, T. SRINIVASAN, S.K. METHA AND R.K. JAIN
Solid State Physics Laboratory, Timarpur, Delhi 110 045, India

F. EICHHORN, R. GROETZSCHEL AND N. SCHELL
Institute for Ion Beam Physics and Materials Research, FZR Rossendorf, 01314 Dresden, Germany

The strain produced or defects generated in the multilayers depend on the extent of lattice mismatch and the layer thickness. Ion irradiation can also induce strain in an otherwise lattice-matched samples. The strain present in strained samples can be tuned using the ion beam mixing methods. The strains have direct bearing on the bandgap engineering in the semiconductor devices. The use of ion beam techniques like RBS/Channeling give reliable information on crystalline quality and quantity of strains and /or defects generated due to any relaxation of strains due to increased layer thickness or increased lattice mismatch. The importance and usefulness of ion beam based techniques to measure and tune strain in such samples is discussed and highlighted in this paper along with our resent results in this direction.

1. Introduction

The objective of this paper is to discuss and highlight the importance of ion beam based methods to measure and engineer the strain and other parameters in semiconductor multilayers. Recent results obtained on the effects of Swift Heavy Ion (*SHI*) irradiation in few such structures will also be presented. Semiconductor superlattices have potential device applications [1-4] for high performance detectors, high speed and high frequency digital and analog circuits

[c] Presenting and corresponding author

because they offer precise control over the states and motions of charge carriers. Band structure of these materials can be tailored to get required electronic and optoelectronic properties. These are basically multilayers with different bandgaps on either side of each interface, and therefore called as *hetero-structures*. The band structure of these materials depends on the band structure of each layer and the band offsets at each interface. Such structures made of III-V binary, ternary, or tertiary compound semiconductors have more applications in the optoelectronics because of their direct bandgap nature. The quantum confinement effects present in these thin multilayer structures opened up doors for a new and exciting technology called semiconductor quantum well (QW) lasers. The bandgap of these materials can be chosen by selectively choosing the composition of different constituent elements [3,4]. All the layers of a superlattice will have similar crystal structure and in general, same lattice parameter. Multilayers with a small lattice mismatch leading to a tensile or compressive strain in the alternating layers are called Strained Layer Superlattice (*SLS*). With the advent of epitaxial growth techniques like Molecular Beam Epitaxy (*MBE*) and Metal Organic Vapor Phase Epitaxy (*MOCVD*) it is now possible to grow crystals with monolayer precision. Atoms deposited on a substrate take positions corresponding to the potential minima of the lattice sites. Hence in the strained-layer epitaxy, despite the difference in substrate and deposited lattice parameters, deposit atoms are constrained to the substrate interatomic spacing in the plane of the interface. Corresponding change occurs in the perpendicular lattice parameter due to the Poisson effect. Both these lattice matched and coherently grown strained systems have their own applications. The strain in the epilayer due to such tetragonal distortion improves the device performance [1-4] and is a parameter for tailoring the band structure and other relevant quantities. For example the strain lifts the degeneracy in the light hole and heavy hole bands at k=0 point of the valence band and thereby reduces the threshold current density required for lasing action in a QW laser. The thickness and composition of the epilayer decides the bandgap of SLS, which can be tuned by the ion irradiation [5-7]. Spatial bandgap tuning of these materials is necessary to integrate the optoelectronic circuits because different devices require different bandgaps. To tune such important parameters of these structures, post growth techniques like ion beam mixing are found to be superior to the manipulation during the growth process [5]. In the following sections we will give a detailed review of the ion beam mixing in superlattice structures, spatial bandgap tuning and the characterization techniques like High Resolution X-Ray Diffraction (*HRXRD*) and RBS/Channeling. Then we will give some of our recent results obtained using SHI irradiation induced mixing.

1.1. *Ion beam mixing in superlattice structures*

As mentioned earlier the integration of optoelectronic devices needs different bandgaps to be accommodated on a single wafer because photonics circuits demand different bandgaps for different devices. For example if we have to integrate a laser, waveguide and any other optical device like a photodetector then we need to choose different bandgap for waveguide so as to reduce the absorption loss. Moreover, most of the time optical devices require alignment to within fractions of a micron and they should be stable under the operational conditions. Hence efficient circuits cannot be fabricated using discrete optical components. So spatial bandgap tuning of heterostructures and multilayers is important. Meeting such bandgap requirements is quite difficult during the growth. The alternative way then would be to alter the bandgap after growing the structures. Compositional disordering and mixing at the interface by Ion Implantation and subsequent thermal annealing is normally employed [5-7]. An ion beam of a given fluence can cause intermixing of atomic layers at the interface, a phenomenon called *Ion Beam Mixing*.

Ion beam mixing has become simple and promising technique because the diffusion/migration of different elements is enhanced by the large concentration gradients exist at each interface. Although these structures are inherently metastable due to these concentration gradients, they are shown to be quite stable under normal operation even at elevated temperatures like 1000o C [8]. Increase in temperature beyond a point say critical temperature tc (900o C for InGaAs based structures) causes the intermixing, but it cannot have any spatial selectivity. The value of this tc can be reduced to a lower value say t1c (around 600o C for InGaAs based structures) by creating some defects like vacancies and interstitials which enhance the diffusion property. Ion irradiation or implantation is found to be more suitable for introducing such defects. Suitable masks can be used to selectively implant the defects over specified area on a wafer and then the temperature of this wafer can be elevated to t1c so as to get different bandgap on these selected spots while the bandgap in the other areas is unchanged. Lot of study has been performed to obtain this planar technology.

Xia et. al [8] gave a detailed review of the ion beam mixing of semiconductor superlattices along with several suitable characterization techniques including HRXRD. They gave an example [9,10] in which AlGaAs alloys were made at the selected areas on an AlAs/GaAs superlattice using Zn ion implantation and subsequent thermal annealing. It is interesting to note that the AlGaAs alloy stays as a single crystal and epitaxial to the surrounding superlattice regions. Since this discovery, impurity disordering in the temperature range of 600o C - 900o C has been reported in many III-V compound semiconductor systems [11-16]. Charabonneu et. al [5] have shown that the blue shift occurs in the QW bandgap of a QW laser due to ion irradiation and subsequent thermal annealing. InGaAs/GaAs and InGaAs/InP based laser structures have been studied using the photoluminescence (PL) as a diagnostic tool. The PL peak position was shown to be shifted due to the mixing effects. The broadening and distortion of PL spectrum of high fluence sample (8.56 MeV As irradiation at 2.5 x 1014 ions/cm2) shows the structural deterioration of the QW laser. Hence different bandgaps can be obtained without loss of the sample quality by keeping the fluence below this value. The well thickness of the irradiated sample was also observed to be increased, using the cross-sectional TEM. In another experiment reported in 1999 [17], they have directly shown the shift in the emission wavelength of an ion mixed QW laser. The bandgap-tuned laser was characterized in terms of threshold current density and external quantum efficiency. They exhibit blue shift in the lasing spectra up to 63nm. This allows laser with a range of tuned wavelengths to be fabricated from a single wafer. Glodberg et. al [18] reported similar experiments in 1997. They have shown the emission spectra of four laser diodes produced from a single semiconductor wafer. The effects of multiple implants on the mixing of AlAs/GaAs have been studied by Venkatesan et. al. [19] using TEM and SIMS, to form AlGaAs alloy layers. The influence of ion mass and fluence on the intermixing of GaAs/AlGaAs quantum well structures using PL and SIMS has been studied in detail by H. Leier et. al. [20]. Diffusion length for Al was measured and compared with the existing theory. Many experiments have been performed to study the ion beam mixing in semiconductor multilayers in this low energy region.

We study the effects SHI irradiation on different proprieties of these materials. This SHI induced mixing, rather than the implantation and subsequent thermal annealing, is more suitable for the integration of optoelectronic devices because of the advantage that the interface mixing can be confined to a narrow region at the interface as against the lateral straggling effects in low energy ion implantation. Moreover SHI mixing studies attain importance from the fundamental need of understanding the energy transfer mechanism in this energy region. The high-energy beam looses its energy to the electronic subsystem of the target material. Then this energy will be transferred to the lattice atoms via the electron phonon coupling. There are two models namely "Coulomb explosion model [21]" and the "Thermal spike model [22]" to explain such energy transfer mechanisms. According the Coulomb explosion model, a positive ion core along the ionizing path of the energetic beam repels the nearby atoms. This causes a radial motion of the atoms sitting around this core. This model is applicable for the insulators like polymers. Thermal spike model assumes that the inelastic energy loss of the energetic beam produces a very high temperature, which exists for a very short time, thereby called as thermal spike. During the spike period the material will melt because these temperatures are very well above the melting temperatures. Diffusion occurs in this molten state but it cannot come back to the equilibrium because of the rapid quenching. Thermal spike is shown to be more responsible for the mixing occurring at the metal-metal and metal - semiconductor interfaces. SHI induced mixing has been recently studied by several authors [23,24]. The basic thrust of such work is to understand the above mentioned energy transfer mechanism as well as to make new materials for novel device applications. Very few such studies have been reported in SLS. Here we study the effects of 130 - 200 MeV Ag ion irradiation on the In.53Ga.47As/InP based lattice-matched and strained InGaAs/GaAs single layers / superlattice structure by HRXRD and RBS/Channeling.

1.2. *RBS and ion channeling*

Rutherford Backscattering Spectroscopy *(RBS)* [25-27] is a powerful tool to determine the thickness, composition, defect densities and strains. When the incident ion beam is directed along a high symmetry crystal direction, then because of the ordered structure of the target atoms, such a particle will undergo a correlated series of small angle gentle collisions, leading to what is known as ion channeling [25-35]. The ranges of channeled particles are enormously large and yield of close impact parameter processes like RBS will be reduced drastically. Hence channeling measurements can be made, by measuring the incident angle dependence of such yields. These angle Vs yield plots, known as angular scans, can be used to map the crystal planes and axis. Angular scans define many structural properties and deformations of the crystal. The amount of reduction of such yield gives a measure of the crystalline quality. The use of ion channeling in conjunction with RBS provides a measure of the crystalline quality as a function of depth. Channeling is sensitive to any lattice defects and has been extensively used to study many types of defects existing in crystals [25-30]. Strain measurements by ion - channeling technique in multilayered structures are based on the tetragonal distortion induced in the layers, which results in the shift ($\Delta\theta$) of the dip of the channeling angular scan measured around off-normal axis. The two components of strain are defined as

$$\varepsilon_{\perp,\parallel} = \frac{a^{\perp,\parallel} - a}{a} = \frac{\Delta a^{\perp,\parallel}}{a} \tag{1}$$

Where a is the lattice constant of that particular layer and superscripts distinguish between the parallel and perpendicular lattice constants. Then the tetragonal distortion ε_t is defined as $\varepsilon_{\|} - \varepsilon_{\perp}$. The off-normal axis (<110> for <100> growth), being parallel to the diagonal of the cubic cell makes an angle θ with the sample plane. This angle can be defined in terms of lattice parameters as $\tan\theta = a_{\perp}/a_{\|}$. Any small change in these lattice parameters due to the strain causes a change in θ, which can be obtained by differentiating the above equation on both sides.

$$\sec^2\theta d\theta = \frac{a_{\|}da_{\perp} - a_{\perp}da_{\|}}{a_{\|}^2} = \frac{a_{\perp}}{a_{\|}}\varepsilon_t = \tan\theta\varepsilon_t \tag{2}$$

$$\Rightarrow \varepsilon_t = \frac{\Delta\theta}{\sin\theta\cos\theta} \tag{3}$$

Hence, one can estimate the strain value by mapping the off normal axis using the channeling techniques. The difference between the positions of dips obtained from layer and the substrate around off normal axis gives the input $\Delta\theta$ for the above equation. The FWHM of this curve is directly related to the channeling critical angle, which varies inversely as square root of incident energy. Hence the angular scans will be sharp in high energy channeling there by the strain resolution and the sensitivity of the method are improved [31-33]. If the energy is too low then the results are mislead by steering effect due to the broad critical angles [33-35]. Hence high-energy channeling measurements are recommended for strain measurements.

1.3. *High Resolution XRD*

HRXRD is an accurate and non-destructive technique [33 & 36] for characterizing epitaxial films, multi layers and the interdiffusion. The strain present in the system modifies the interplanar spacing, hence HRXRD is sensitive to lattice strain. In optimum conditions it is capable of measuring the strain with a sensitivity of about 10^{-5}. For example Figs. 1&2, show the HRXRD patterns measured from multi layers and Fig. 3 corresponds to single strained layer. The sharp and intense peak in the center corresponding to Bragg angle (for the substrate material and for particular reflection) is known as the substrate peak. The width of the peak has the information about the crystalline quality as well as the angular resolution of the measurement. A broad peak appearing on the left side to the substrate peak in fig. 3 corresponds to the layer i.e the thin strained layer grown on GaAs substrate and hence is called as layer peak. The tetragonal distortion due to the presence of strain in this layer, modifies the interplanar spacing and thereby the Bragg reflection angle. Hence the layer peak got separated from the substrate peak. It appears on the left side for the compressive strain and on the right side for the tensile strain. The peak is broad and the intensity is low because it corresponds to a thin layer. The layer thickness and the composition can be estimated using the simulation programs. Figs. 1 & 2 show the patterns obtained for multilayers. The system of equi-spaced low intensity peaks around the substrate peak is due to the interference between the reflections emerging from different layers. These are called satellite peaks, a good number of satellite orders appear if the interfaces are sufficiently sharp. Here the center of the satellite system matches with the substrate peak for lattice matched samples. Hence the shift ($\Delta\theta$) in the layer peak position (or the center of the satellite system) with respect to that of the substrate peak is a measure of the strain ($\varepsilon_\perp$). It can be easily estimated by differentiating the Bragg's law ($2d \sin\theta = n\lambda$), as $-\Delta\theta\cot\theta$. One can determine the value of $\varepsilon_{||}$ either by using the crystal elastic constants combined with Poisson ratio or by perform the HRXRD along other asymmetric directions. One can also use the knowledge of ε_τ obtained from RBS/Channeling to calculate $\varepsilon_{||}$. The superlattice period (*s*) of multilayers can be calculated from the interference pattern. If θ_1, θ_2 are the angular positions of two consecutive satellite orders then $s = \lambda/2(\sin\theta_1 - \sin\theta_2)$. The reduction of the satellite intensities after irradiation represents the interdiffusion. The measured strain, composition and thickness help in understanding the mixing effects. Hence HRXRD is a suitable technique for characterizing these materials and also to study the mixing effects.

2. Experimental

Necessary samples were grown using MBE facility at Solid State Physics Laboratory, Delhi and the MOCVD facility at Warsaw, Poland. Sample specifications are provided in table 1. Then each of these samples was cut in to two counter parts with one part kept unirradiated. The other part was irradiated by energetic Ag ions delivered from 16MV Pelletron accelerator at Nuclear Science Centre, New Delhi. *Ag* ($\geq$ 130 MeV) was selected for being above S_e (electronic energy loss) threshold for both *InP* and *GaAs*. For example, S_e threshold for InP is around 13 KeV/nm while the S_e of 130 MeV Ag in InP is around 14 KeV/nm. The irradiation details are also given in table 1. The sample identity (ID) given in this table will be used to refer any particular sample. The letter "U" at the end of the ID indicates that it is Unirradiated, whereas the letter "I" indicates that it is irradiated. Pristine and irradiated P523 were again cut for performing rapid thermal annealing (RTA). These samples have been annealed at 450° C for 90 sec. in N_2 atmosphere using RTA facility at Dresden, Germany.

Annealed samples are referred with a tag "RTA" at the end of the corresponding sample's name. Samples have been characterized by HRXRD near the reflection <004> (GaAs) (for MQW5) <006> (InP) (for P523) at ROBL Material Research Station, Grenoble using the synchrotron radiation. The incident x-ray wavelength was 0.153nm. The other samples were characterized by HRXRD near <004> (GaAS) reflection at SSPL, Delhi, using the X'Pert Materials research diffractormeter. RBS channeling studies on P523 were reported earlier [37]

Table 1: Sample specifications and irradiation details

Sample ID	*Specification*	*Growth technique*	*Beam*	*Fluence (ions/cm^2)*
2601	$In_{0.1}Ga_{0.9}As$(250Å)/GaAs	MBE	150 MeV Ag	1 x 10^{13}
4101	$In_{0.1}Ga_{0.9}As$(400 Å)/GaAs	"	150 MeV Ag	1 x 10^{13}
4401	$In_{0.14}Ga_{0.47}As$ (75Å) / GaAs (~200 Å) ...x 10 /GaAs	"	200 MeV Ag	5 x 10^{12}
MQW5	$In_{0.12}Ga_{0.88}As$ (75 Å) / GaAs (150 Å) ....x 20 /GaAs	"	130 MeV Ag	5 x 10^{12}
P523	$In_{0.53}Ga_{0.47}As$ (150 Å) /InP(150 Å)......x 10 /InP	MOCVD	130 MeV Ag	5 x 10^{12}

3. Results and discussion

Figs. 1-3 show the HRXRD spectra of MQW5, P523 and 4101 respectively. Spectra of other samples are not shown here. Each figure contains all irradiated, unirradiated and / or annealed spectra of a particular sample so that it will be easy for comparison. These spectra are displaced (Shifted upwards) on the intensity axis for clarity. Present HRXRD results and the earlier reported channeling results [37] suggest that the samples are of considerably good crystalline and interface quality. It is also observed that the irradiation did not spoil these structures significantly. The measured superlattice periods of all the multilayer samples are found to be lesser than that of the nominal values as shown in table 2. Measured strain values are also shown in table 2.

MWQ5 is basically strained (compressive) one, if we assume the pseudomorphic growth, then the perpendicular lattice strain of 1.65% is calculated (for 12% *In*) using Ve-gard law, whereas the inplane lattice strain is zero. With this value the strain averaged over the multilayer amounts to be

0.55%. However the strain values obtained from the HRXRD analysis are around 0.3%. Exact values are given in the table 2. This low strain value along with the low satellite intensities indicates the less *In* composition in the layers than that of the nominal (12%) value. *In* composition corresponding to the measured strain values are 7%and 6% for MWQ5U and MQW5I respectively.

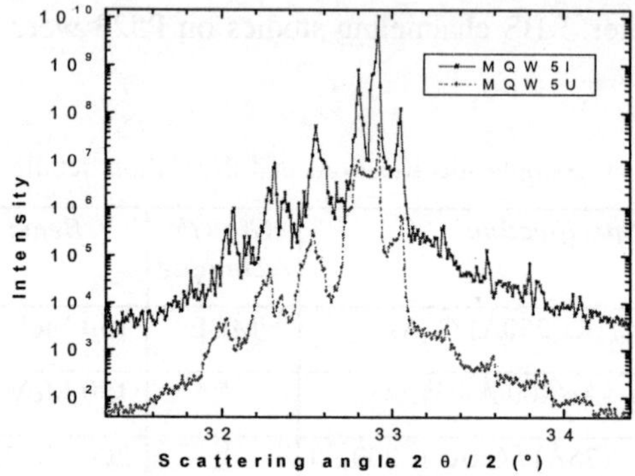

Fig. 1: HRXRD spectra of MQW5U and MQW5I

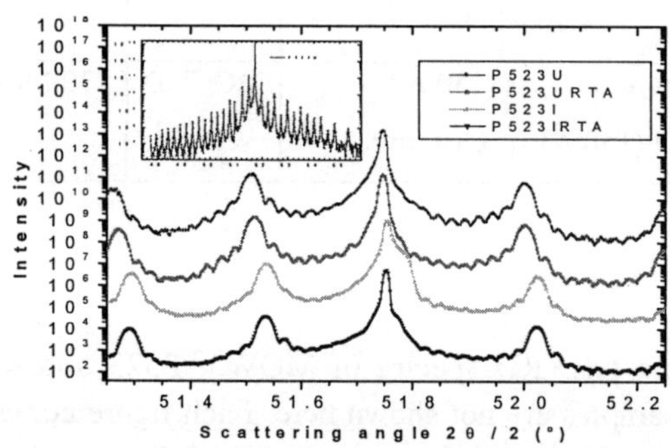

Figure 2: HRXRD spectra of P523 series samples

From fig.1 it is also observed that there is a secondary system existing in addition to the main system of superlattice with slightly less *In* concentration (4% for MQW5U and 3% for MQW5I) in some layers. However the ratio of such layers to the main system is very small (1:3). Lower satellite orders in HRXRD suggest that the interfaces may not be very sharp. The reduction of *In* concentration represented by the reduction of the satellite intensities, in the irradiated sample reflects the gradual diffusion of *In* from InGaAs layers. Correspondingly there is a reduction of strain from 0.325% to 0.298%. Although the measured periods of both the samples are found to be less than the nominal value, there is a $10A^0$ difference between MWQ5U and MWQ5I. The superlattice period of the irradiated sample is found to be $10A^0$ larger that that of pristine sample. The other InGaAs/GaAs based systems also show similar behavior. The reduction in the compressive strains can also be seen in table 2.

Table 2 HRXRD results

S. No	Sample	S (nm)	$\varepsilon_\perp$%		*S. No*	*Sample*	$\varepsilon_\perp$%
1	MQW5U	20.15	0.325		7	2601	0.959
2	MQW5I	21.18	0.298		8	2601I	0.943
3	P523 U	28.90	-0.0110		9	4101	1.010
4	P523 URTA	29.08	-0.0144		10	4101I	0.897
5	P523 I	29.01	-0.0346		11	4401	0.271
6	P523 IRTA	28.93	-0.0359		12	4401I	0.192

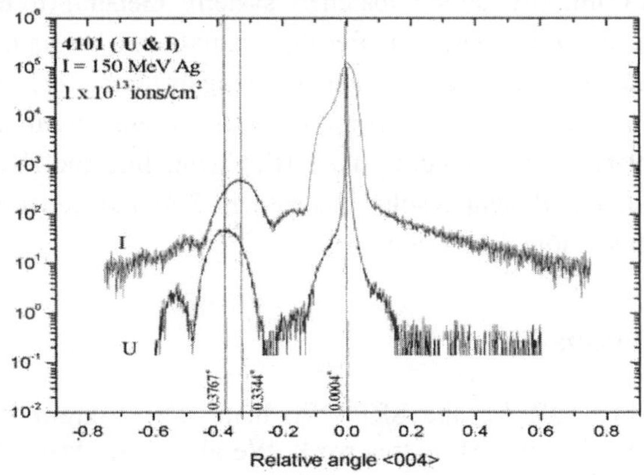

Fig.3: HRXRD spectra of 4101 U and 4101I

SHI irradiation induces a tensile strain in an initially lattice matched $In_{0.53}Ga_{0.47}As/InP$ based multilayer system i.e P523. Very high number of satellite orders in the HRXRD spectra shown in fig.2, indicate that the interfaces are very sharp and the boundaries are almost rectangular in shape. Also indicated is the fact that these interfaces remain considerably sharp even after irradiation. The period of the irradiated sample is found to be less than that of the unirradiated sample. Similar feature is observed in InGaAs/GaAs systems also. The mixing effects are more prominent for ion beam processed and annealed sample. Simple irradiation changes the strain but the interface quality is regained after annealing. However annealing alone doesn't change the strain value significantly. RBS channeling experiments performed on this sample were reported earlier [37]. A good reduction in the channeling spectra of P523U and P523I show a good crystalline quality. Channeling angular scans indicate that there exists a finite strain in the irradiated sample which is measured accurately using HRXRD in the present work. The gradual diffusion of *In* from the surface and the migration of *Ga* or *As* like atoms to surface has been explicitly observed in the RBS spectrum. Such modification of strain in all the above-mentioned samples, using the SHI mixing is the main motivation of the present work.

4. Conclusions

The general trend in all the samples indicates the gradual diffusion of *In* from surface and the migration of *Ga or As* like atoms to the surface regions due to the SHI irradiation and/or annealing processes. The compressive strain is found to decrease in the initially compressive strained samples and tensile strain is induced in an initially lattice matched system. General trend indicates the increase in the superlattice period after the irradiation. The importance of the ion beam based methods to obtain the spatial bandgap tuning has been discussed in detail. A brief review of the existing work in low energy mixing has also been given. Characterization techniques like RBS/Channeling and HRXRD have been discussed in detail. Recent results obtained in SHI induced mixing have been presented with suitable discussion.

Acknowledgements

SVSNR thanks CSIR for the SRF. We thank Dr. Vikram Kumar, Director, SSPL, Delhi, for his interest in this work. We also thank Prof. A. Turos, IEMT, Warsaw for providing one of the samples (P523). Part of the work was supported by a research project (No: ERIP/ER/0103296/M/01) sponsored by DRDO, India.

References

1. M.S. Shur, T.A. Fjeldly and A.D. Bykhovski, J. Appl. Phys., 85, 3009, 1999.
2. T. Ashely, C.T. Elliott, N.T. Gordon, R.S.Hall, A.D.Johnson and G.J.Pryce, Appl. Phys. Lett., 64, 118, 1994.
3. S.T. Picraux, B.L. Doyle and J.Y. Tsao in Strained-layer Superlattices: Materials Science and Technology, Vol 33 "Semiconductors and Semimetals", ed: Thomas P Pearsall, Academic Press, New York, 1991.

4. G.C. Osbourn, J. Appl. Phys., 53, 1586, 1982.
5. S. Charabonneu, P.J. Poole, P.G. Pive, M. Buchanan, R.D. Goldberg and I.V. Mitchell, Nucl. Inst. and Meth. (B) 106, 457, 1995.
6. J.W. Wan, D.A. Thompson and J.G. Simmons, Nucl. Instr. Meth. B, 106, 461, 1995.
7. V. Hofsaβ, J. Kuhn, C. Kaden, V. Harle, H. Bolay, F. Scholz, H. Schweizer, H. Hillmer, R. Losch and W. Schlapp, Nucl. Instr. Meth. B, 106, 471, 1995.
8. W. Xia, S.N. Hsu, C.C. Han, S.A. Papert, B.Zhu, C. Cozzolino, P.K.L. Yu and S.S. Lau, Nucl. Instrum. Meth. B, 59/60, 491, 1991.
9. W.D. Laidig, N. Holonyak Jr., M.D. Camras, K. Hess, J.J. Coleman, P.D. Dapkus and J. Bardeen, Appl. Phys. Lett., 38, 776, 1981.
10. N. Hplonyak Jr., W.D. Laidig, M.D. Camras, J.J. Coleman and P.D. Dapkus, Appl. Phys. Lett., 39, 776, 1981
11. D. Deppe and N. Holnyak Jr., J. Appl. Phys., 64, R93, 1988
12. K.H. Lee, H.H. Park and D.A. Stevenson, J. Appl. Phys., 65, 1048, 1989
13. J.M. Dallesasse, W.E. Plano, D.W. Nam, K.C. Hsieh, J.E. Baker and N. Hplonyak Jr., J. Appl. Phys 66, 482, 1989.
14. W.D. Laidig, J.W. Lee, P.K. Chiang, L.W. Simpson and S.M. Bedair, J. Appl. Phys. 54, 6382, 1983.
15. B. Tell, J. Shah, P.M. Thomas, K.F. Brown-Goebeler, A. DiGiovanni, B.I. Miller and U. Koren, Appl. Phys. Lett. 54, 1570, 1989.
16. S. O'Brien, D.P. Bour and R.J. Shealy, Appl. Phys. Lett. 53, 1859, 1988
17. S. Charabonneu, P.J. Poole, Y. Feng, G.C. Aers, M. Dion, M.Davies R.D. Goldberg and I.V. Mitchell, Appl. Phys. Plett., 67, 2954, 1995
18. R.D. Goldberg, I.V. Mitchell, P. Poole, D. Labrie, H. Lafontaine, G.C. Aers, R. Williams, M. Dion, S. Charbonneau, K. Ramanujancha and G.C. Weatherly, Nucl. Instrum. Meth. B., 127/128, 418, 1997.
19. T. Venkatesan, S.A. Schwarz, D.M. Hwang, R. Bhat, H.W. Yoon and Y. Arakawa, Nucl. Instrum. Meth. B., 19/20, 777, 1987.
20. H. Leier, A. Forchel, G. Horcher, J. Hommel, S. Bayer, H. Rothfritz, G. Weimann and W. Schlapp., J. Appl. Phys., 67, 1805, 1990.
21. R.L. Fleisher, P.B. Prile and R.M. Walker, Nuclear tracks in solids, Univ. of California Press, California, 1975.
22. Z. G. Wang, Ch. Duforur, E. Paumier and M. Toulemonde, J. Phys. C, 6, 6733, 194
23. W. Assmann, M. Dobler, D.K. Avasthi, S.Kruiser, H.D. Mieskes, H. Nolte, Nucl. Instrum. Meth. B, 146, 271, 1998.
24. C. Dufor, Ph. Bauer, G. Marchal, J. Grilhe, C. Jaouen, J. Pacaud, J.C. Jousset, Euro Phys. Lett., 21, 671, 1993.
25. W.K. Chu, J.W. Mayer and M.A. Nicolet, Backingscattering Spectrometry, Academic Press, New York, 1978.
26. L.C. Feldman and J.W. Mayer, Fundamentals of Surface and Thin film Analysis, North-Holland, New York, 1986

27. L.C. Feldman, J.W. Mayer and S.T. Picraux, Material Analysis by Ion Channeling, Acadmic Press, New York, 1982.
28. J. Lindhard, Mat. Fys. Medd. Dan. Vid. Selsk. 34, 1, 1965.
29. A. P. Pathak, Radiation Effects and defects, 30, 193, 1976.
30. D.S. Gemmel, Rev. of Mod. Phys.46,129, 1974.
31. D.K. Avasthi, S.V.S. Nageswara Rao and A.P. Pathak, Proce. DAE-BRNS workshop on thin films and multilayers, 68-71, oct. 1999.
32. H. Nolte, W. Assmann, H. Huber, S. A. Karamian and H. D. Mieskes, Nucl. Instrum. Meth. B, 136, 587, 1998
33. A. M. Siddiqui, S.V.S. Nageswara Rao, A. P. Pathak, V. N. Kulkarni, R. Kesav Murthy, E. Williams, Daryush Ila, Claudiu Muntele, K.S. Chandrasekaran and B. M. Arora, J. Appl. Phys. 90, 2824, 2001.
34. A.P. Pathak, S.V.S. Nageswara Rao, A. M. Siddiqui, Nucl. Inst. and Meth. (B), 161-163, 487, 2000.
35. Shin Hashimoto, Y.Q. Feng and W.M. Gibson, Nucl. Inst. and Meth. B, 13, 45, 1986.
36. J.H.C. Hogg, D. Shaw and D.M. Staudte, Appl. Surf. Sc., 50, 87, 1991.
37. A.P. Pathak, S.V.S. Nageswara Rao, A.M. Siddiqui, G.B.V.S. Lakshmi, S.K. Srivastava, S. Ghosh, D. Bhattacharya, D.K. Avasthi, D.K. Goswami, P. V. Satyam, B.N. Dev and A. Turos, Nucl. Insrt. Meth. B, 193, 319, 20042.

SURFACE MODIFICATIONS IN SILICON (100) DUE TO ANTIMONY IMPLANTATION

SHIKHA VARMA, SOMA DEY
Institute of Physics,
Bhubaneswar - 751005, India.
E-mail: shikha@iopb.res.in

V. GANESAN
IUC-DAEF,
University Campus,
Khandwa Road,
Indore, India

The modifications in surface roughness after 1.5 MeV Sb ion implantation have been investigated by the Atomic Force Microscopy. The surface roughness demonstrates three varied behaviours as a function of fluence. An initial low roughness and slow roughening regime gives way to a higher and faster increase in roughness beyond a critical fluence. Finally, a saturation of surface roughness is observed. Surface roughening, thus unexpectedly occurs in 3 stages, governed by the fluence, and increases in a non-monotonous fashion. We have further investigated the lattice disorder, created after ion implantation, by utilizing the technique of Rutherford Back Scattering/Channeling. A Multiple Scattering Formalism has been applied to extract the disorder density from the Rutherford Back Scattering/Channeling data. The disorder density also shows three trends as a function of fluence. Surprisingly, we observe a simple linear relationship between the lattice disorder and the surface roughness up to the fluence of 1×10^{14} $ions/cm^2$. Results demonstrate that though the surface is primarily determined by surface effects, lattice damage may also play a significant role in surface roughness.

1. Introduction

In Si based semiconductor technology Sb is considered an important dopant for its role in the development of FETs and infrared detectors [1]. MeV ion implantation plays a crucial role in semiconductor device technology, and is therefore widely utilized in many device fabrications. Demands for higher density and faster devices have led the microelectronics industry to continued miniaturization of device structures. This trend has introduced new issues that must be considered and one of these is surface roughness. Roughness can determine the performance and reliability of devices [2] and therefore, it becomes necessary to be able to characterize it. Although there are some studies that investigate evolution of surface roughness after keV ion implantation, such studies after MeV implantation have received very little attention [3]. Moreover there are some suggestions that lattice defects and strains, produced via implantation, may also play role in determining the surface evolution [4]. These effects are however poorly understood. In the present study we investigate the surface roughness as well as the lattice disorder created in the silicon sample after MeV implantation. We find that though the surface effects will predominantly determine the surface, lattice effects e.g. radiation induced damage etc. may also play a significant role.

In this study, we have utilized a combination of Atomic Force Microscopy (AFM) and Rutherford Back Scattering/ Channeling (RBS/C) techniques to measure the surface roughness and the lattice disorder, respectively, in Si after implantation. RBS/C data has also been utilized to extract the depth dependent disorder in Si lattice. In the present study we find that MeV Sb ion implantation in Si leads to surface roughnesses that display three different trends as a function of fluence. We further find a direct - linear relationship between the surface roughness and the disorder in the Si-lattice up to an Sb fluence of 1×10^{14} $ions/cm^2$. These results have not been observed earlier and can be of immense importance for the fundamental understanding as well as modeling and fabrication of the desired surfaces. After discussing the experimental procedures in section 2, we study roughness of the Si(100) surfaces in section 3. The disorder in Si lattice and its relationship with the surface roughness is also examined.

2. Experimental

Si(100) samples were implanted at room temperature with a beam of 1.5 MeV Sb^{2+} ions at various fluences ranging from 1×10^{12} to 5×10^{15} $ions/cm^2$. AFM Nanoscope E was used to image the silicon sample surfaces both prior to and after implantation. RBS/C measurements were carried out along the [100] axis, using 2.5 MeV He^{+2} ions at a scattering angle of 150°.

3. Results And Discussion

Antimony ions of 1.5 MeV were implanted in silicon and AFM was utilized to acquire images of the ion beam modified surfaces. About 6-8 surface images of $1 \times 1\ \mu m^2$ were acquired for virgin (unimplanted) Si sample as well as for Si samples after implantation at each fluence. The root mean square (rms) roughness of all these surfaces was measured and the average roughness at each fluence is plotted in Fig. 1. Surface roughness exhibits three prominent behaviors as a function of fluence. For low fluences, up to $1 \times 10^{13}\ ions/cm^2$, the roughness is small and does not increase much compared to the virgin surface roughness. Beyond this fluence, an enhanced surface roughness, increasing at a much steeper rate is observed. This trend continues up to the fluence of $1 \times 10^{14}\ ions/cm^2$ where a high roughness of 0.296 nm is measured. A saturation in surface roughness is observed beyond this fluence. The observed effects indicate that the surface roughening is not a continuous process and occurs in three distinct steps. This anomalous behavior has never been reported earlier.

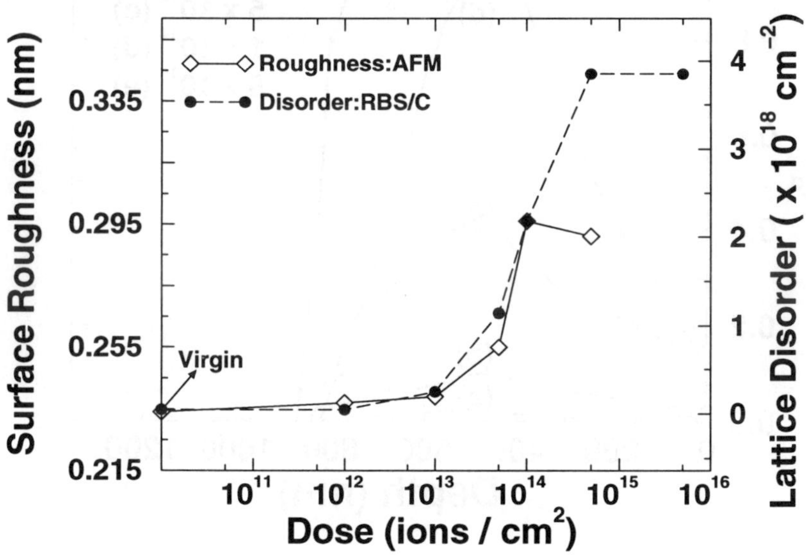

Figure 1. The rms roughness of the Sb implanted Si(100) surfaces and the density of lattice disorder in Si after implantation (see text) as a function of fluence. Symbol sizes denote the error in the measurement. Data for the virgin sample is also shown.

The depth dependent disorder density in Si lattice, created by Sb ion implantation is shown in Fig. 2. This has been extracted by applying the multiple scattering formalism, described by Feldman and Rodgers [5] to the RBS/C data (not shown). The details of parameters used for these calculation have been described elsewhere [6]. The ordinate is the ratio of the density of the displaced atoms N_d and the atomic density of host silicon atoms N. Small lattice disorder is noticed for fluences of 1×10^{12} $ions/cm^2$ and 1×10^{13} $ions/cm^2$. A damage peak at 460 nm below the surface and a buried amorphous layer are observed at 5×10^{13} $ions/cm^2$ and 1×10^{14} $ions/cm^2$, respectively. After 5×10^{14} $ions/cm^2$, the top 900 nm of Si lattice, including the surface layer, becomes amorphous. Our earlier studies of MeV Sb implantations in Si have shown that this damage can be reduced by high temperature anneals ($\sim$ 600^{0}C to 800^{0}C) [7,8] and high energy C^+ irradiation [9].

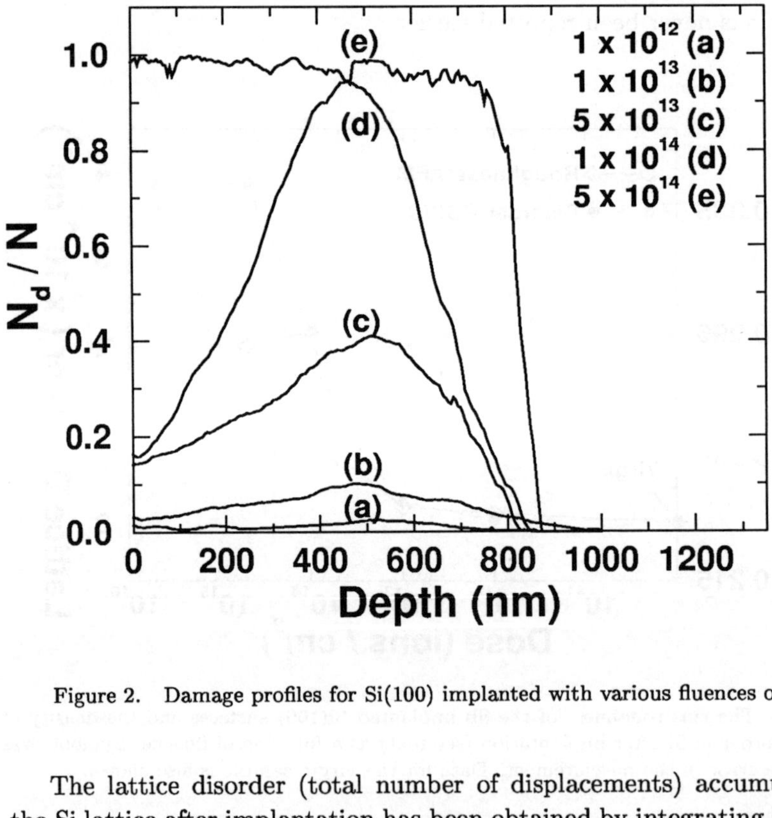

Figure 2. Damage profiles for Si(100) implanted with various fluences of Sb.

The lattice disorder (total number of displacements) accumulated in the Si lattice after implantation has been obtained by integrating N_d/N (of Fig. 2) and is plotted in Fig. 1 as a function of Sb fluence. Surprisingly, as seen in Fig. 1, the lattice disorder displays three trends very similar to those demonstrated by surface roughness as a function of fluence. Initial phase is of low lattice-disorder, low surface roughness and their small increase up to 1×10^{13} $ions/cm^2$. This is taken over by a stage where much larger lattice-disorder and a higher surface roughness are observed upto 1×10^{14} $ions/cm^2$. During this period, both lattice disorder and surface roughness increase at a rate which is much faster than that of the previous stage. Finally, a saturation of surface roughness at 1×10^{14} $ions/cm^2$ and of lattice-disorder at 5×10^{14} $ions/cm^2$ is observed. Our Raman results have shown that the first signatures of amorphicity in the Si lattice after 1.5 MeV Sb ion implantation occurs at the critical fluence, D_c, 1×10^{13} $ions/cm^2$ [8]. In Si lattice, point defects will be the dominant defect species at fluences lower than D_c. With simple point defects, which are isolated and randomly distributed in Si-matrix, the total lattice disorder will be small. However, once the defects become complex, with the initiation of amorphicity in the Si- lattice at D_c, the lattice disorder increases rapidly. Lattice disorder and surface roughness saturate at different fluences owing to the fact that disorder saturates only when the thick amorphous layer approaches the surface (see Fig. 2) where as the saturation in surface roughness will be related to critical number of damaged zones on the surface. As seen in fig. 1, in addition to the similar behaviors and trends, a one to one correspondence, up to fluence of 1×10^{14} $ions/cm^2$, is found between the data of surface roughness and the lattice disorder. These results are unexpected and effectively demonstrate that surface roughness, up to the fluence of 1×10^{14} $ions/cm^2$, exhibits a direct- linear relationship with the total damage in the Si-lattice. Details of this relationship have been studied elsewhere [10]. The linear relationship between the total accumulated disorder and the surface roughness, up to 1×10^{14} $ions/cm^2$, fluence where surface roughness gets saturated, has never been suggested earlier and can have significant implications in MeV implantation related technologies. Results suggest that in addition to surface effects, lattice disorder also plays a role in determining the surface roughness.

4. Conclusions

In the present study the modifications in the surface roughness of the Si(100) surfaces have been examined after 1.5 MeV Sb implantation. Techniques of AFM and RBS/C have been utilized in conjunction to investigate the relationship between the surface roughness and the disorder created during implantation. Surface roughness exhibits three varied trends as a function of Sb fluence indicating that surface roughening is not a simple monotonous process. Furthermore, we find that for fluences less than 1×10^{14} $ions/cm^2$, surface roughness displays a simple linear relationship with the total accumulated lattice disorder.

Acknowledgments

This work is partly supported by ONR grant no. N00014-97-1-0991.

References

1. G. Tempel, N. Schwarz, F. Muller, F. Koch, H.P. Zeindl, and I. Eisele, *Thin Solid Films* **174**, 171 (1990).
2. D.J. DiMaria and D.R. Kerr, *Appl. Phys. Lett.* **27**, 505 (1975).
3. G. Carter, M.J. Nobes, I.V. Katardjiev and J.L. Whitton *Defect and Diffusion Forum*, **vol.57/58** (1988) 97-126. Ion Implantation 1988, ed. F.H. Wohlbier (Trans. Techn. Publ. Ltd).
4. L. Porte, M. Phaner, C.H. de Villeneuve, N. Moncoffre, and J. Tousset, *Nucl. Instr. Meth.* **B44**, 116 (1989).
5. L.C. Feldman and J.W. Rodgers, *J. Appl. Phys.* **41**, 3776 (1970).
6. S. Dey, A. Pradhan and S. Varma, *J. Vac. Sci. Tech.* **B18**, 2457 (2000).
7. S. Dey, G. Kuri, B. Rout and S. Varma, *Nucl. Instr. Meth.* **B142**, 35 (1998).
8. S. Dey, C. Roy, A. Pradhan and S. Varma, *J. Appl. Phys.* **87**, 1110 (2000).
9. S. Dey, and S. Varma, *J. Appl. Phys.* **90**, 2618 (2001).
10. S. Varma, S. Dey and V. Ganesan, submitted to *J. Appl. Phys.*.

Author Index